FREE Test Taking Tips DVD Offer

To help us better serve you, we have developed a Test Taking Tips DVD that we would like to give you for FREE. **This DVD covers world-class test taking tips that you can use to be even more successful when you are taking your test.**

All that we ask is that you email us your feedback about your study guide. Please let us know what you thought about it – whether that is good, bad or indifferent.

To get your **FREE Test Taking Tips DVD**, email freedvd@studyguideteam.com with "FREE DVD" in the subject line and the following information in the body of the email:

a. The title of your study guide.

b. Your product rating on a scale of 1-5, with 5 being the highest rating.

c. Your feedback about the study guide. What did you think of it?

d. Your full name and shipping address to send your free DVD.

If you have any questions or concerns, please don't hesitate to contact us at freedvd@studyguideteam.com.

Thanks again!

SIFT Study Guide 2020 and 2021

SIFT Test Study Guide 2020-2021 and Practice Exam Questions for the Military Flight Aptitude Test [4th Edition]

Test Prep Books

Copyright © 2020 Test Prep Books

All rights reserved.

Table of Contents

Quick Overview

As you draw closer to taking your exam, effective preparation becomes more and more important. Thankfully, you have this study guide to help you get ready. Use this guide to help keep your studying on track and refer to it often.

This study guide contains several key sections that will help you be successful on your exam. The guide contains tips for what you should do the night before and the day of the test. Also included are test-taking tips. Knowing the right information is not always enough. Many well-prepared test takers struggle with exams. These tips will help equip you to accurately read, assess, and answer test questions.

A large part of the guide is devoted to showing you what content to expect on the exam and to helping you better understand that content. In this guide are practice test questions so that you can see how well you have grasped the content. Then, answer explanations are provided so that you can understand why you missed certain questions.

Don't try to cram the night before you take your exam. This is not a wise strategy for a few reasons. First, your retention of the information will be low. Your time would be better used by reviewing information you already know rather than trying to learn a lot of new information. Second, you will likely become stressed as you try to gain a large amount of knowledge in a short amount of time. Third, you will be depriving yourself of sleep. So be sure to go to bed at a reasonable time the night before. Being well-rested helps you focus and remain calm.

Be sure to eat a substantial breakfast the morning of the exam. If you are taking the exam in the afternoon, be sure to have a good lunch as well. Being hungry is distracting and can make it difficult to focus. You have hopefully spent lots of time preparing for the exam. Don't let an empty stomach get in the way of success!

When travelling to the testing center, leave earlier than needed. That way, you have a buffer in case you experience any delays. This will help you remain calm and will keep you from missing your appointment time at the testing center.

Be sure to pace yourself during the exam. Don't try to rush through the exam. There is no need to risk performing poorly on the exam just so you can leave the testing center early. Allow yourself to use all of the allotted time if needed.

Remain positive while taking the exam even if you feel like you are performing poorly. Thinking about the content you should have mastered will not help you perform better on the exam.

Once the exam is complete, take some time to relax. Even if you feel that you need to take the exam again, you will be well served by some down time before you begin studying again. It's often easier to convince yourself to study if you know that it will come with a reward!

Test-Taking Strategies

1. Predicting the Answer

When you feel confident in your preparation for a multiple-choice test, try predicting the answer before reading the answer choices. This is especially useful on questions that test objective factual knowledge. By predicting the answer before reading the available choices, you eliminate the possibility that you will be distracted or led astray by an incorrect answer choice. You will feel more confident in your selection if you read the question, predict the answer, and then find your prediction among the answer choices. After using this strategy, be sure to still read all of the answer choices carefully and completely. If you feel unprepared, you should not attempt to predict the answers. This would be a waste of time and an opportunity for your mind to wander in the wrong direction.

2. Reading the Whole Question

Too often, test takers scan a multiple-choice question, recognize a few familiar words, and immediately jump to the answer choices. Test authors are aware of this common impatience, and they will sometimes prey upon it. For instance, a test author might subtly turn the question into a negative, or he or she might redirect the focus of the question right at the end. The only way to avoid falling into these traps is to read the entirety of the question carefully before reading the answer choices.

3. Looking for Wrong Answers

Long and complicated multiple-choice questions can be intimidating. One way to simplify a difficult multiple-choice question is to eliminate all of the answer choices that are clearly wrong. In most sets of answers, there will be at least one selection that can be dismissed right away. If the test is administered on paper, the test taker could draw a line through it to indicate that it may be ignored; otherwise, the test taker will have to perform this operation mentally or on scratch paper. In either case, once the obviously incorrect answers have been eliminated, the remaining choices may be considered. Sometimes identifying the clearly wrong answers will give the test taker some information about the correct answer. For instance, if one of the remaining answer choices is a direct opposite of one of the eliminated answer choices, it may well be the correct answer. The opposite of obviously wrong is obviously right! Of course, this is not always the case. Some answers are obviously incorrect simply because they are irrelevant to the question being asked. Still, identifying and eliminating some incorrect answer choices is a good way to simplify a multiple-choice question.

4. Don't Overanalyze

Anxious test takers often overanalyze questions. When you are nervous, your brain will often run wild, causing you to make associations and discover clues that don't actually exist. If you feel that this may be a problem for you, do whatever you can to slow down during the test. Try taking a deep breath or counting to ten. As you read and consider the question, restrict yourself to the particular words used by the author. Avoid thought tangents about what the author *really* meant, or what he or she was *trying* to say. The only things that matter on a multiple-choice test are the words that are actually in the question. You must avoid reading too much into a multiple-choice question, or supposing that the writer meant something other than what he or she wrote.

5. No Need for Panic

It is wise to learn as many strategies as possible before taking a multiple-choice test, but it is likely that you will come across a few questions for which you simply don't know the answer. In this situation, avoid panicking. Because most multiple-choice tests include dozens of questions, the relative value of a single wrong answer is small. As much as possible, you should compartmentalize each question on a multiple-choice test. In other words, you should not allow your feelings about one question to affect your success on the others. When you find a question that you either don't understand or don't know how to answer, just take a deep breath and do your best. Read the entire question slowly and carefully. Try rephrasing the question a couple of different ways. Then, read all of the answer choices carefully. After eliminating obviously wrong answers, make a selection and move on to the next question.

6. Confusing Answer Choices

When working on a difficult multiple-choice question, there may be a tendency to focus on the answer choices that are the easiest to understand. Many people, whether consciously or not, gravitate to the answer choices that require the least concentration, knowledge, and memory. This is a mistake. When you come across an answer choice that is confusing, you should give it extra attention. A question might be confusing because you do not know the subject matter to which it refers. If this is the case, don't eliminate the answer before you have affirmatively settled on another. When you come across an answer choice of this type, set it aside as you look at the remaining choices. If you can confidently assert that one of the other choices is correct, you can leave the confusing answer aside. Otherwise, you will need to take a moment to try to better understand the confusing answer choice. Rephrasing is one way to tease out the sense of a confusing answer choice.

7. Your First Instinct

Many people struggle with multiple-choice tests because they overthink the questions. If you have studied sufficiently for the test, you should be prepared to trust your first instinct once you have carefully and completely read the question and all of the answer choices. There is a great deal of research suggesting that the mind can come to the correct conclusion very quickly once it has obtained all of the relevant information. At times, it may seem to you as if your intuition is working faster even than your reasoning mind. This may in fact be true. The knowledge you obtain while studying may be retrieved from your subconscious before you have a chance to work out the associations that support it. Verify your instinct by working out the reasons that it should be trusted.

8. Key Words

Many test takers struggle with multiple-choice questions because they have poor reading comprehension skills. Quickly reading and understanding a multiple-choice question requires a mixture of skill and experience. To help with this, try jotting down a few key words and phrases on a piece of scrap paper. Doing this concentrates the process of reading and forces the mind to weigh the relative importance of the question's parts. In selecting words and phrases to write down, the test taker thinks about the question more deeply and carefully. This is especially true for multiple-choice questions that are preceded by a long prompt.

9. Subtle Negatives

One of the oldest tricks in the multiple-choice test writer's book is to subtly reverse the meaning of a question with a word like *not* or *except*. If you are not paying attention to each word in the question, you can easily be led astray by this trick. For instance, a common question format is, "Which of the following is...?" Obviously, if the question instead is, "Which of the following is not...?," then the answer will be quite different. Even worse, the test makers are aware of the potential for this mistake and will include one answer choice that would be correct if the question were not negated or reversed. A test taker who misses the reversal will find what he or she believes to be a correct answer and will be so confident that he or she will fail to reread the question and discover the original error. The only way to avoid this is to practice a wide variety of multiple-choice questions and to pay close attention to each and every word.

10. Reading Every Answer Choice

It may seem obvious, but you should always read every one of the answer choices! Too many test takers fall into the habit of scanning the question and assuming that they understand the question because they recognize a few key words. From there, they pick the first answer choice that answers the question they believe they have read. Test takers who read all of the answer choices might discover that one of the latter answer choices is actually *more* correct. Moreover, reading all of the answer choices can remind you of facts related to the question that can help you arrive at the correct answer. Sometimes, a misstatement or incorrect detail in one of the latter answer choices will trigger your memory of the subject and will enable you to find the right answer. Failing to read all of the answer choices is like not reading all of the items on a restaurant menu: you might miss out on the perfect choice.

11. Spot the Hedges

One of the keys to success on multiple-choice tests is paying close attention to every word. This is never truer than with words like almost, most, some, and sometimes. These words are called "hedges" because they indicate that a statement is not totally true or not true in every place and time. An absolute statement will contain no hedges, but in many subjects, the answers are not always straightforward or absolute. There are always exceptions to the rules in these subjects. For this reason, you should favor those multiple-choice questions that contain hedging language. The presence of qualifying words indicates that the author is taking special care with his or her words, which is certainly important when composing the right answer. After all, there are many ways to be wrong, but there is only one way to be right! For this reason, it is wise to avoid answers that are absolute when taking a multiple-choice test. An absolute answer is one that says things are either all one way or all another. They often include words like *every*, *always*, *best*, and *never*. If you are taking a multiple-choice test in a subject that doesn't lend itself to absolute answers, be on your guard if you see any of these words.

12. Long Answers

In many subject areas, the answers are not simple. As already mentioned, the right answer often requires hedges. Another common feature of the answers to a complex or subjective question are qualifying clauses, which are groups of words that subtly modify the meaning of the sentence. If the question or answer choice describes a rule to which there are exceptions or the subject matter is complicated, ambiguous, or confusing, the correct answer will require many words in order to be expressed clearly and accurately. In essence, you should not be deterred by answer choices that seem excessively long. Oftentimes, the author of the text will not be able to write the correct answer without offering some qualifications and modifications. Your job is to read the answer choices thoroughly and completely and to select the one that most accurately and precisely answers the question.

13. Restating to Understand

Sometimes, a question on a multiple-choice test is difficult not because of what it asks but because of how it is written. If this is the case, restate the question or answer choice in different words. This process serves a couple of important purposes. First, it forces you to concentrate on the core of the question. In order to rephrase the question accurately, you have to understand it well. Rephrasing the question will concentrate your mind on the key words and ideas. Second, it will present the information to your mind in a fresh way. This process may trigger your memory and render some useful scrap of information picked up while studying.

14. True Statements

Sometimes an answer choice will be true in itself, but it does not answer the question. This is one of the main reasons why it is essential to read the question carefully and completely before proceeding to the answer choices. Too often, test takers skip ahead to the answer choices and look for true statements. Having found one of these, they are content to select it without reference to the question above. Obviously, this provides an easy way for test makers to play tricks. The savvy test taker will always read the entire question before turning to the answer choices. Then, having settled on a correct answer choice, he or she will refer to the original question and ensure that the selected answer is relevant. The mistake of choosing a correct-but-irrelevant answer choice is especially common on questions related to specific pieces of objective knowledge. A prepared test taker will have a wealth of factual knowledge at his or her disposal, and should not be careless in its application.

15. No Patterns

One of the more dangerous ideas that circulates about multiple-choice tests is that the correct answers tend to fall into patterns. These erroneous ideas range from a belief that B and C are the most common right answers, to the idea that an unprepared test-taker should answer "A-B-A-C-A-D-A-B-A." It cannot be emphasized enough that pattern-seeking of this type is exactly the WRONG way to approach a multiple-choice test. To begin with, it is highly unlikely that the test maker will plot the correct answers according to some predetermined pattern. The questions are scrambled and delivered in a random order. Furthermore, even if the test maker was following a pattern in the assignation of correct answers, there is no reason why the test taker would know which pattern he or she was using. Any attempt to discern a pattern in the answer choices is a waste of time and a distraction from the real work of taking the test. A test taker would be much better served by extra preparation before the test than by reliance on a pattern in the answers.

FREE DVD OFFER

Don't forget that doing well on your exam includes both understanding the test content and understanding how to use what you know to do well on the test. We offer a completely FREE Test Taking Tips DVD that covers world class test taking tips that you can use to be even more successful when you are taking your test.

All that we ask is that you email us your feedback about your study guide. To get your **FREE Test Taking Tips DVD**, email freedvd@studyguideteam.com with "FREE DVD" in the subject line and the following information in the body of the email:

- The title of your study guide.
- Your product rating on a scale of 1-5, with 5 being the highest rating.
- Your feedback about the study guide. What did you think of it?
- Your full name and shipping address to send your free DVD.

Introduction to the SIFT

Function of the Test

The Selection Instrument for Flight Training (SIFT) exam measures multiple aptitudes and assesses mathematical skills, the ability to extract meaning from passages, mechanical concepts and simple machines, and the ability to recognize patterns within groups of images. Aviation terminology, familiarity with aircrafts and aerodynamics, and knowledge of flight rules and regulations are also assessed. The SIFT exam is used to predict training performance in aviation; generally, individuals who do well on the SIFT will be successful in an aviation training program. Therefore, this test will aid in the selection process, to ensure the most efficient and competent individuals are selected for the Army Aviation Program. The Army is currently checking on the validity of the SIFT, and the Navy and Air Force use similar testing.

Test Administration

Test takers are given three hours to take the SIFT exam. This time includes checking in at the exam site, setup, an optional break, and working through the test itself. Most individuals can complete the exam within 2 hours. The SIFT is not available in paper format and is conducted via computer over a secure server. An individual is not permitted to retake the SIFT if a passing score of any kind is obtained. If the minimum score is not obtained on the first attempt, a second attempt can be made, but no sooner than 181 days after the first attempt. If the individual still does not achieve the minimum passing score on the second attempt, no further attempts can be made, and the individual is no longer qualified for the Army's Aviation Program. Testing sites include Post-Servicing Education Centers, Military Entrance Processing Stations (MEPS), and Military Academies/ROTC programs. Tests need to be scheduled at the location where they will be administered. For additional assistance, an individual may contact his or her local servicing education center or recruiter.

On testing day, candidates must present photo identification and their social security card. Calculators and electronic devices are not permitted but formulas are provided on the test as well as scrap paper. The SIFT is only administered in English and no other special accommodations are currently offered.

Test Format

There are seven categories on the SIFT, each of which requires a variety of skills. Individuals are advised to not randomly guess as time runs out because random guessing is more detrimental to a score than the penalty incurred for unanswered questions. There are time limits for each of the subtests. The table below outlines the number of questions and allotted time for each test section:

Test Section	Time Limit	Section Length
Simple Drawings (SD)	2 minutes	100 questions
Hidden Figures (HF)	5 minutes	50 questions
Army Aviation Information Test (AAIT)	30 minutes	40 questions
Spatial Apperception Test (SAT)	10 minutes	25 questions
Reading Comprehension Test (RCT)	30 minutes	20 questions
Math Skills Test (MST)	40 minutes	Variable length
Mechanical Comprehension Test (MCT)	15 minutes	Variable length

Scoring

To apply for the Army's Aviation program, candidates must currently meet the minimum passing score of 40. Scores range from 20 to 80, with a mean of 50 and standard deviation of 10.

Scores are provided immediately after completing the test. The Test Control Officer (TCO) or Test Examiner (TE) grants the test taker a score letter, which has the TCO's or TE's signature. This signature on the score letter is required in order for the report to be valid. Test takers are advised to strive to perform equally well on each subtest, even though the formula used to compute the SIFT score is proprietary information. A poor score on one subtest may be offset by a higher score on another subtest.

Recent/Future Developments

The Army is currently conducting validity research on the SIFT. Army personnel are assessing the SIFT, as tests like this one are valuable in predicting performance. The current scoring system of the SIFT may change as the validity information for the SIFT is obtained. The minimum score requirement may change to correspond with the Aviation Branch or U.S. Army Recruiting Commands' requirements.

Simple Drawings

The Simple Drawings section of the SIFT provides a unique challenge because, while the problems are individually quite simple, the sheer volume of questions makes the pace of the section rigorous. Each question in this section asks test takers to examine a set of five items and then choose the outlier—the one that does not fit with the set. Essentially, four of the five items are identical and one is a mismatch. Test takers must quickly identify the one item that is different and indicate their selection on the answer sheet.

Each question in this section contains a row with five simple drawings labeled A-E. For example, there may be four identical triangles and a circle; the circle is the outlier. Another question may depict four identical simple arrows and then a single line, missing the arrow head, or four solid shaded circles and one that is only outlined. Again, the correct answer is obvious but speeding through 100 in two minutes, or one question every 1.2 seconds, is nearly impossible. For this reason, test takers should not be discouraged if they do not finish the section. The goal is to get through as many of the 100 questions as possible while maintaining accuracy. Incorrect responses do penalize one's score, so any unanswered questions that remain as time winds down should not be randomly guessed or hastily selected. Although some guesses may be correct, because incorrect responses lower one's score, a better strategy is simply to answer as many as possible that one has had a moment to actually view.

While the Simple Drawings section may seem like an unfair or a pointless task if nearly all test takers are unable to complete the whole set, the Army has intentionally designed the section with this level of simplicity and rigor. The section is able to assess two important aptitudes for hopeful Army pilot candidates—reaction time under pressure and accurate observation under pressure. While taking exams is certainly stressful, being an Army pilot will involve high pressure and high stakes conditions regularly. For this reason, the Army admissions officers use results from sections such as Simple Drawings to efficiently screen the readiness of candidates to handle such situations.

Test takers are encouraged to practice working through Simple Drawings problems prior to encountering this section on the official SIFT exam attempt. The practice questions presented here are very similar to those that will appear on the test. To mimic testing conditions and practice racing the clock, test takers should set a two-minute timer and work through as many questions as possible while trying to maintain accuracy. It is normal to feel stressed by the timer and fatigue towards the end of the two minutes. Anxious test taker should remember that hardly anyone is able to complete all of the problems. However, by staying calm and focused, test takers can strike the optimal balance of speed and accuracy to achieve a high score.

It should be noted that the experience of answering the questions in this section will be somewhat more challenging on the official SIFT administration. In paper form, test takers have the luxury of seeing a full page of several questions at one time, while the questions on the official SIFT administered via computer are presented one at a time. While seemingly minor, this affects the actual amount of time to answer the questions and the resultant pace that must be maintained. When multiple problems are visible at one time, the time to process the information and select a response is augmented because test takers can essentially do several problems in their visual field simultaneously and then simply circle the correct answers on autopilot. On the computer, the question must be processed and answered prior to seeing the next question. Even if it only takes a tenth of a second longer to process each question via computer administration, this time accumulates over the 100 questions to 10 seconds, reducing the effective testing window to 110 seconds. Therefore, to better simulate the authentic test experience, test takers are encouraged to use a blank sheet of paper to cover up the subsequent questions on each page and only focus on one question at a time.

One final note is that the practice questions in this section can, and should, be attempted numerous times. As mentioned, the problems in the Simple Drawings section involve rapidly comparing groups of random figures. As such, it is unlikely that test takers will retain much of the information they process during practice runs, unlike when retaking science practice questions where answers can be memorized. To benefit from additional practice trials, it is recommended that test takers mark answers on a separate piece of paper.

Practice Questions

1.

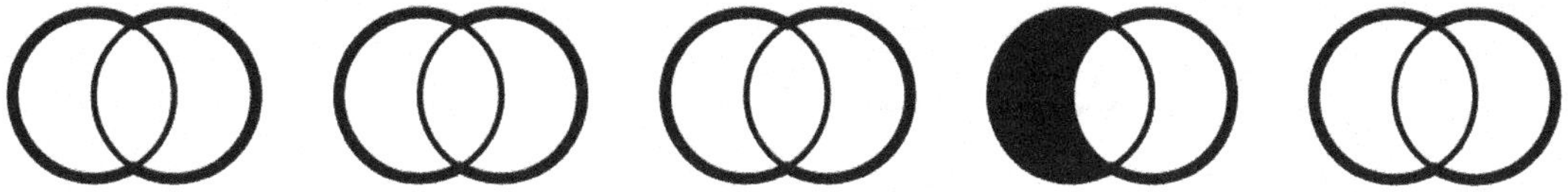

A B C D E

2.

A B C D E

3.

A B C D E

4.

A B C D E

5.

A B C D E

6.

A B C D E

7.

A B C D E

8.

A B C D E

9.

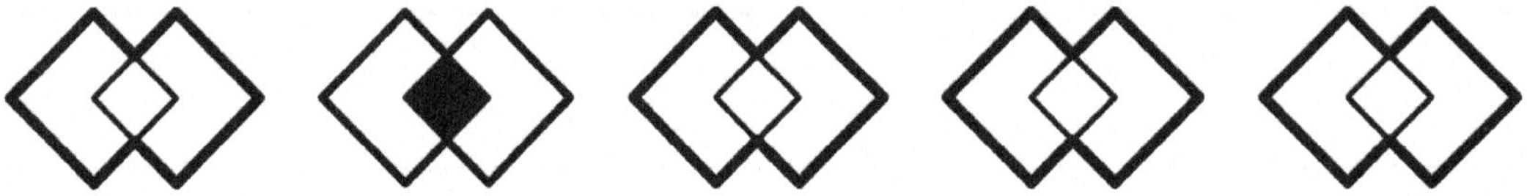

A B C D E

10.

A B C D E

11.

A B C D E

12.

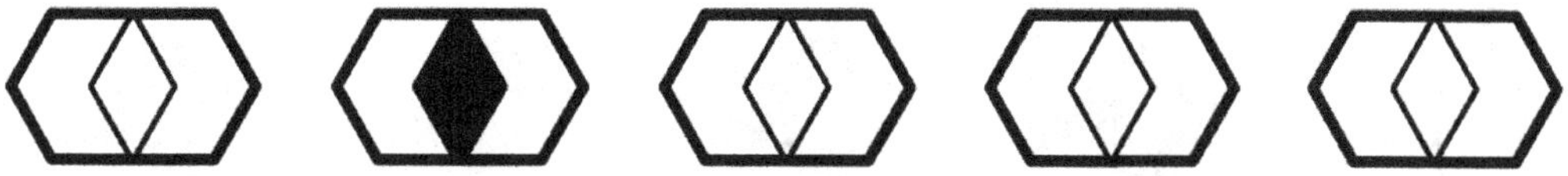

A B C D E

13.

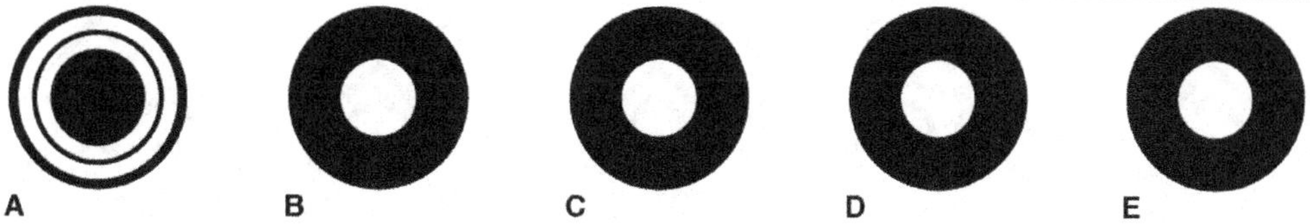

A B C D E

14.

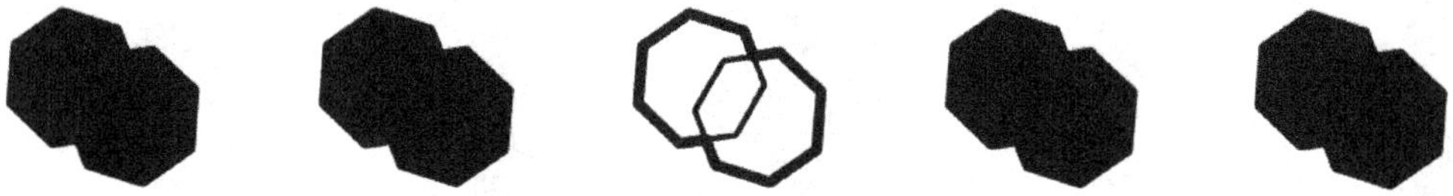

A B C D E

15.

A B C D E

16.

A B C D E

17.

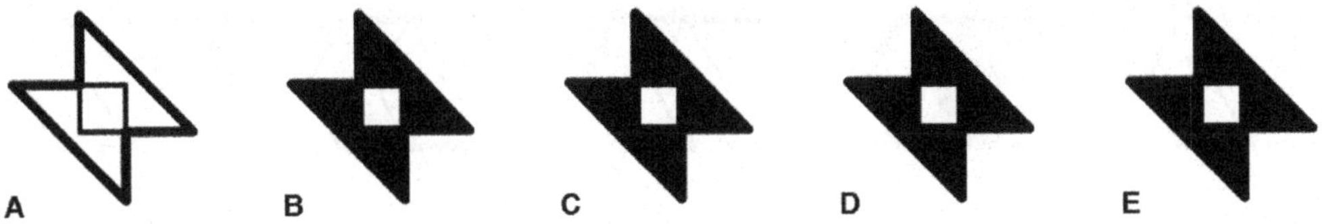

A B C D E

18.

A B C D E

19.

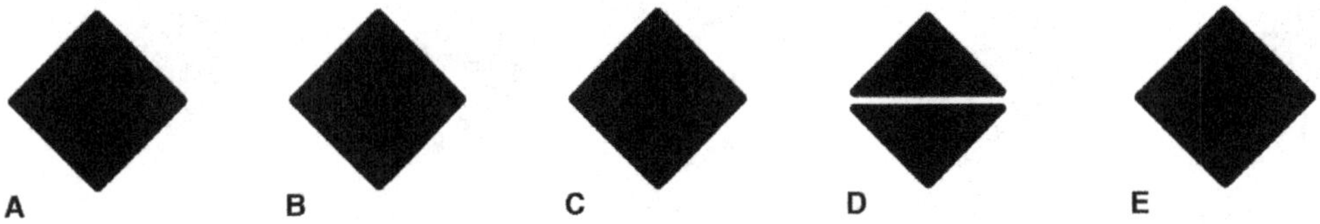

A B C D E

20.

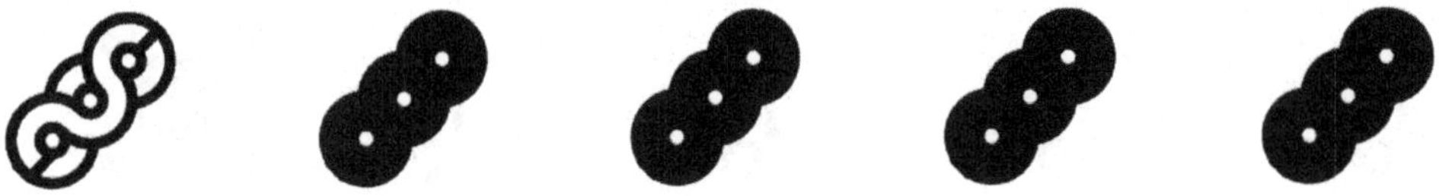

A B C D E

21.

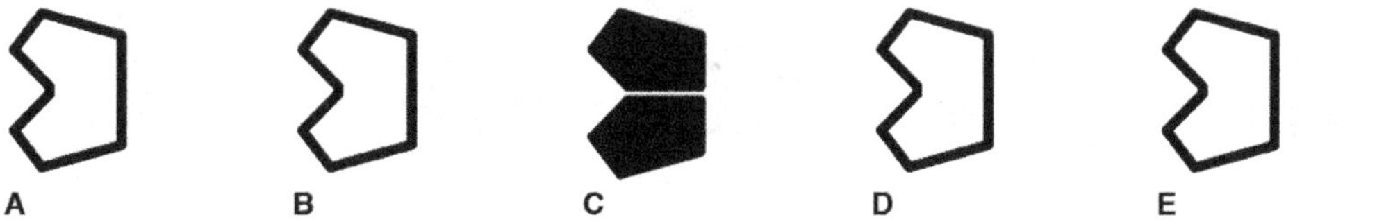

22.

23.

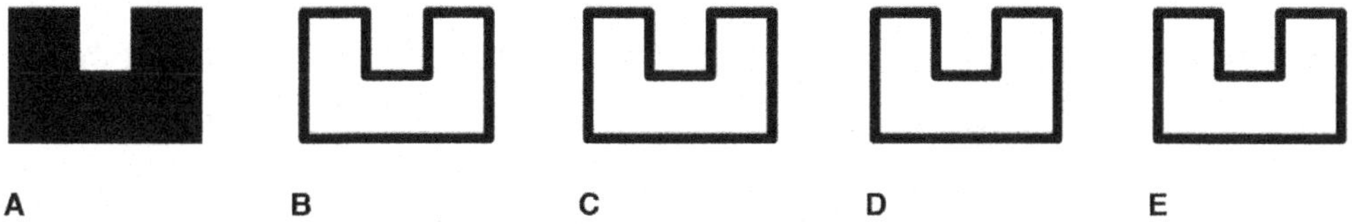

24.

25.

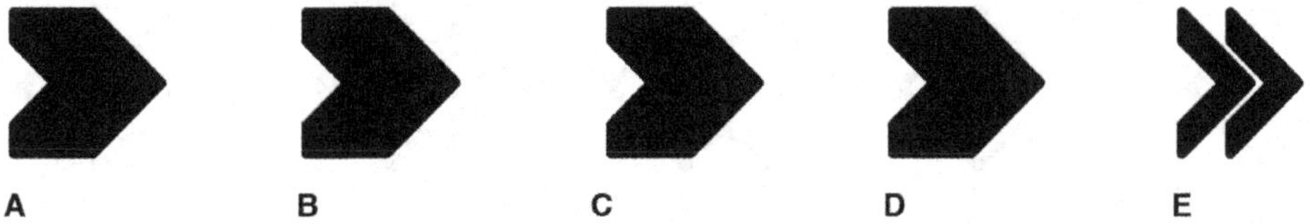

26.

A B C D E

27.

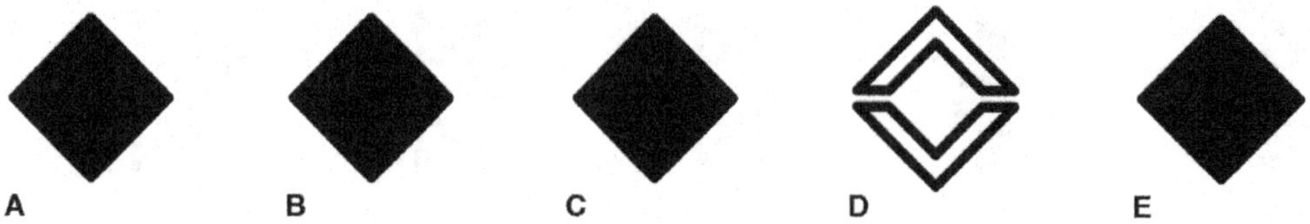

A B C D E

28.

A B C D E

29.

A B C D E

30.

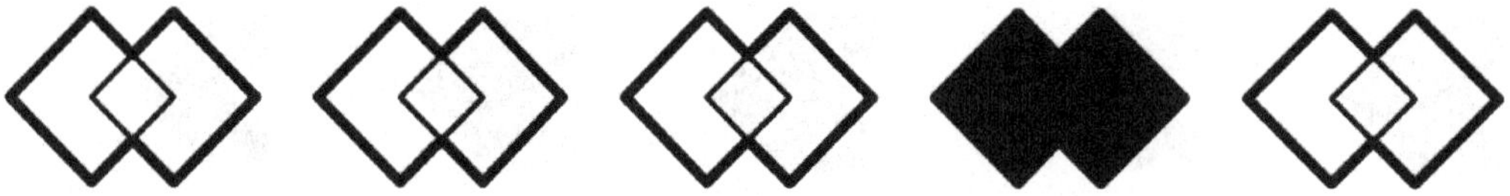

A B C D E

31.

32.

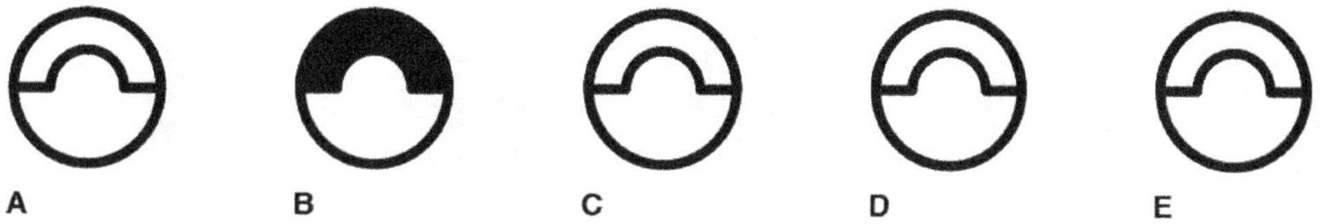

33.

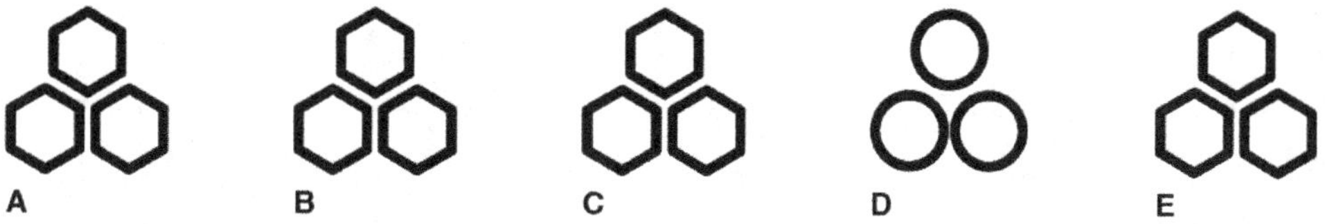

34.

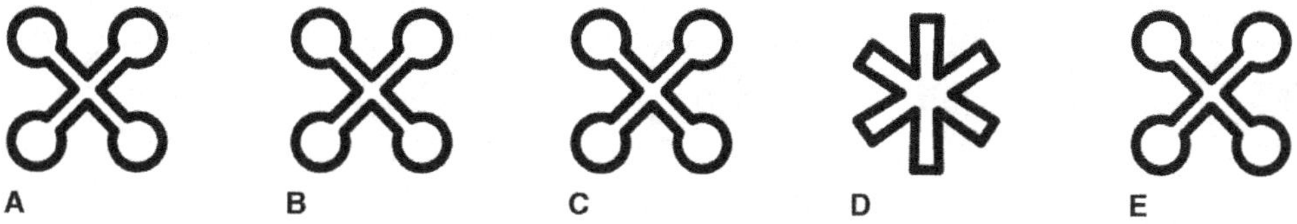

35.

36.

A B C D E

37.

A B C D E

38.

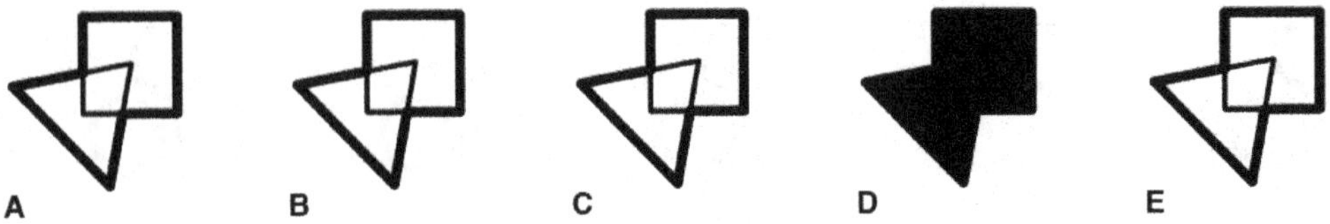

A B C D E

39.

A B C D E

40.

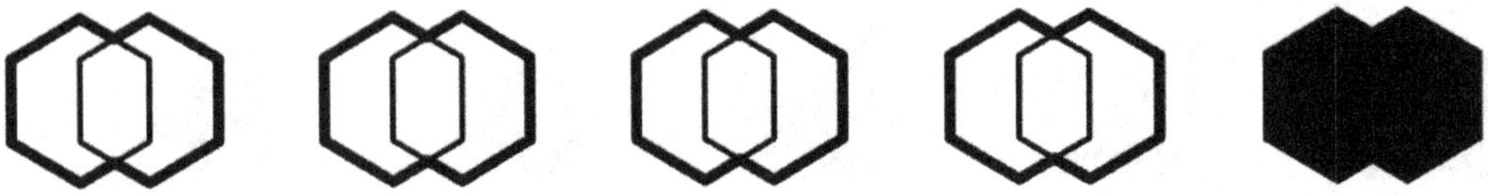

A B C D E

41.

A B C D E

42.

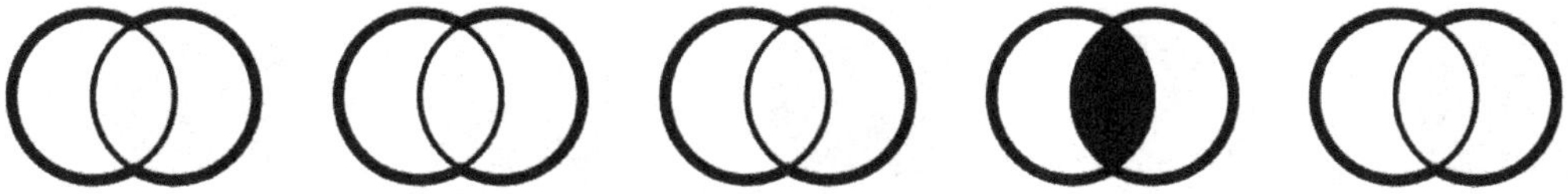

A B C D E

43.

A B C D E

44.

A B C D E

45.

A B C D E

46.

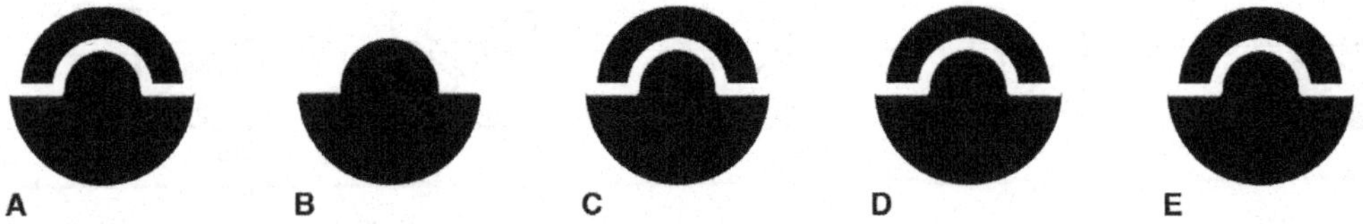

47.

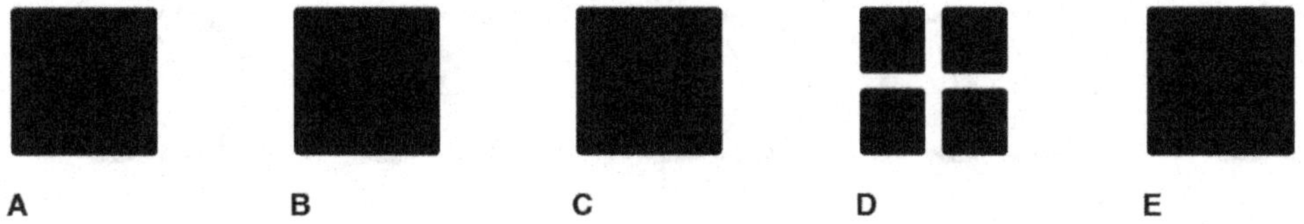

48.

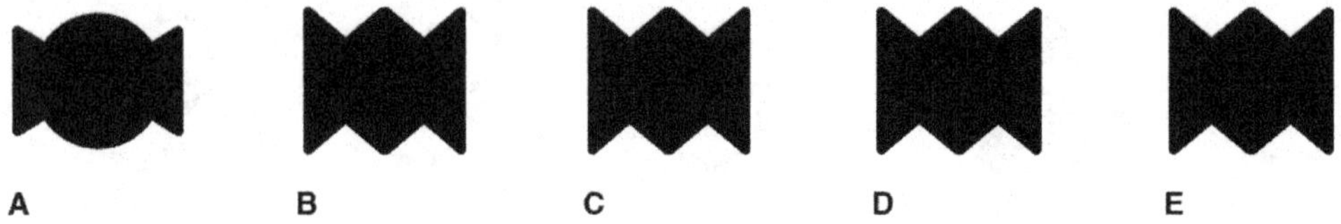

49.

50.

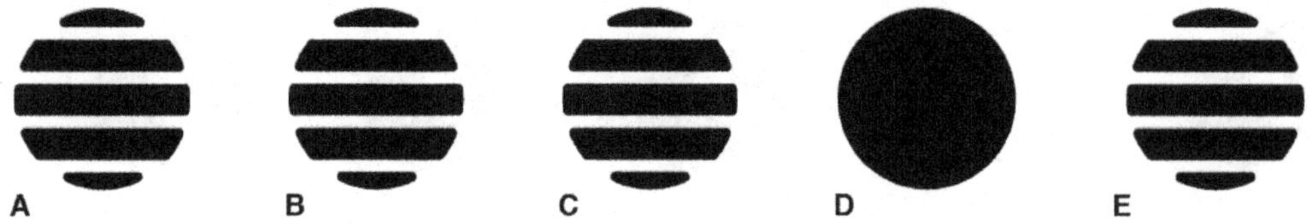

51.

A B C D E

52.

A B C D E

53.

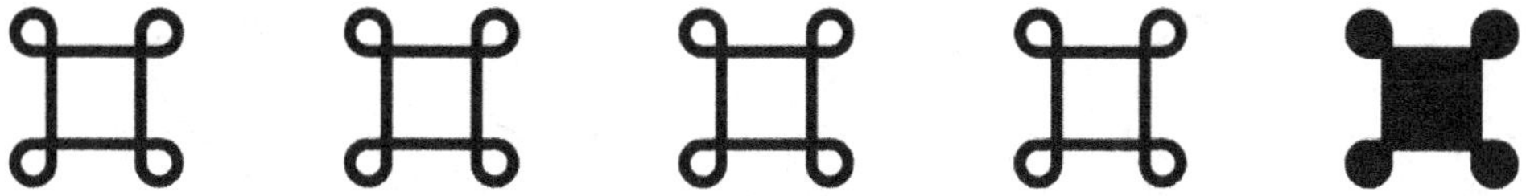

A B C D E

54.

A B C D E

55.

A B C D E

56.

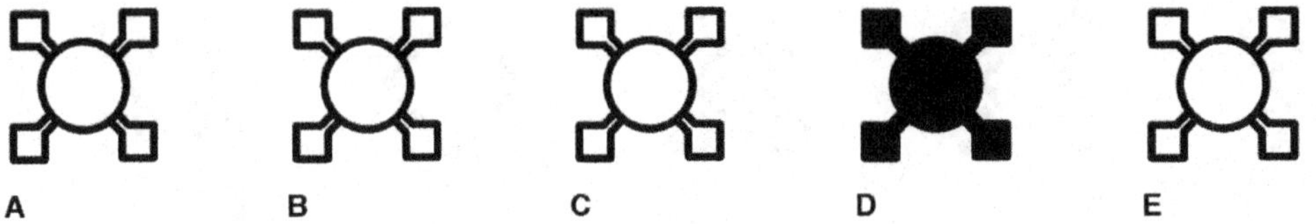

57.

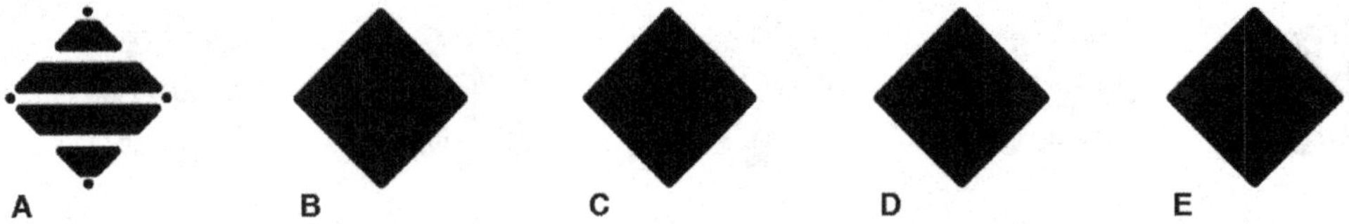

58.

59.

60.

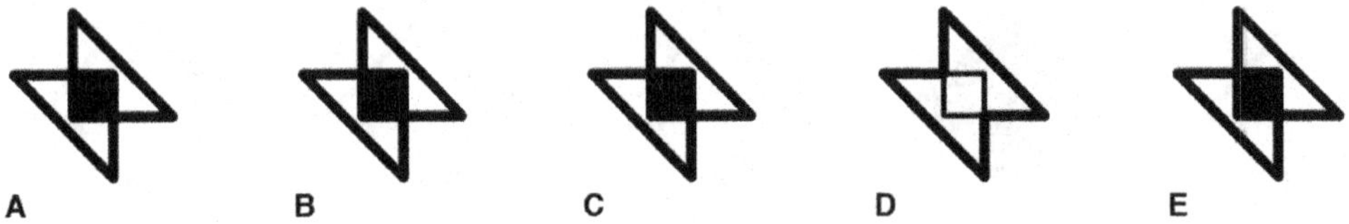

61.

A B C D E

62.

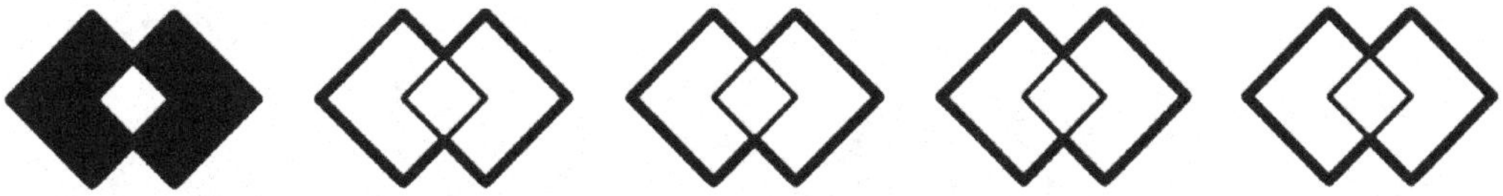

A B C D E

63.

A B C D E

64.

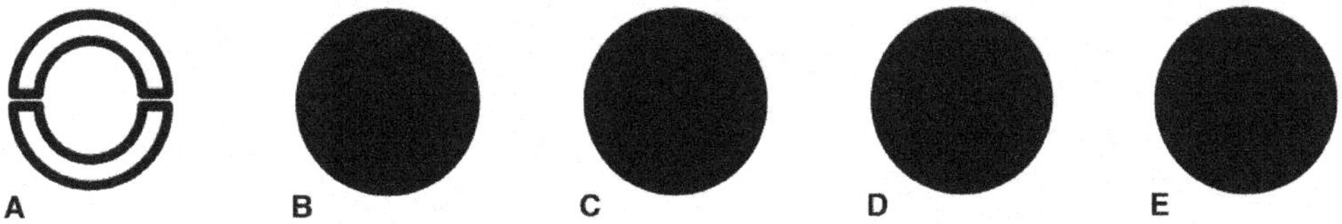

A B C D E

65.

A B C D E

66.

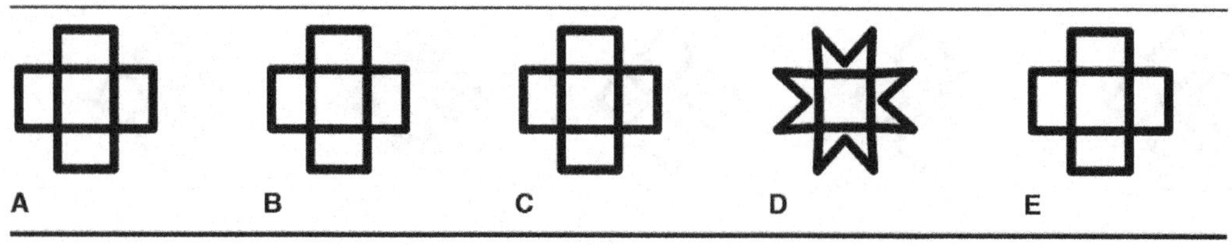

67.

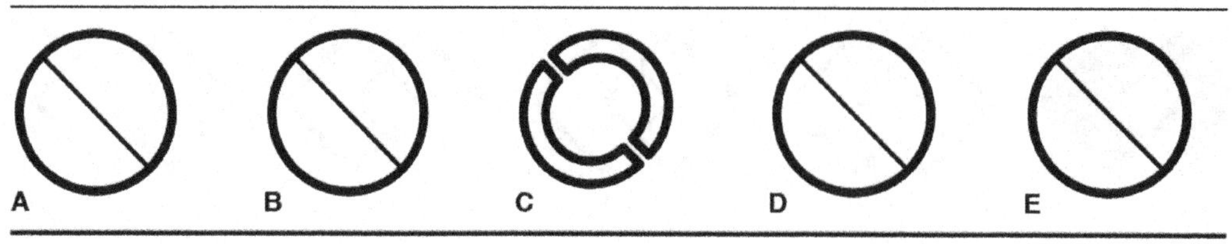

68.

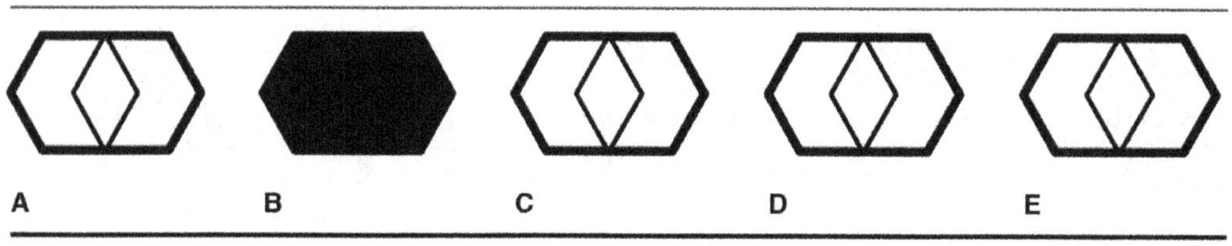

69.

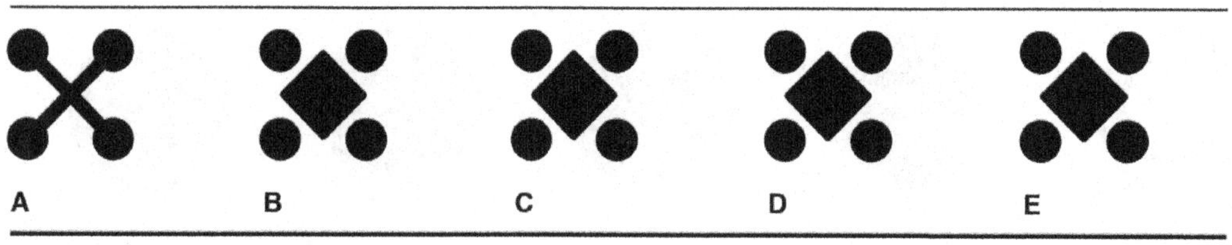

70.

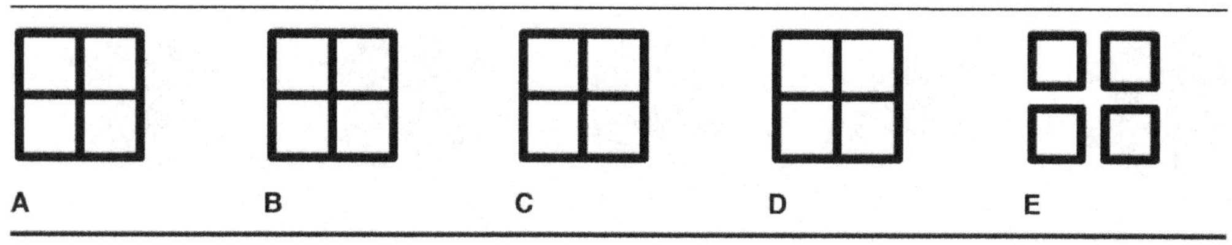

71.

A B C D E

72.

A B C D E

73.

A B C D E

74.

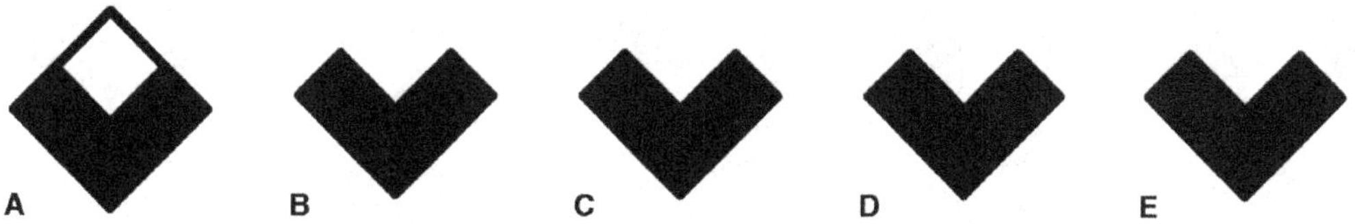

A B C D E

75.

A B C D E

76.

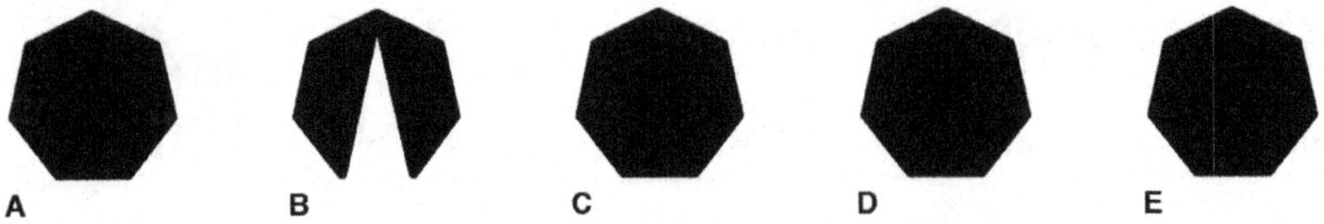

77.

78.

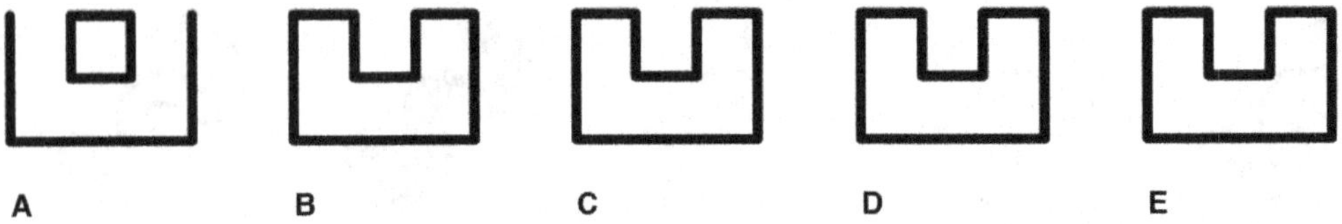

79.

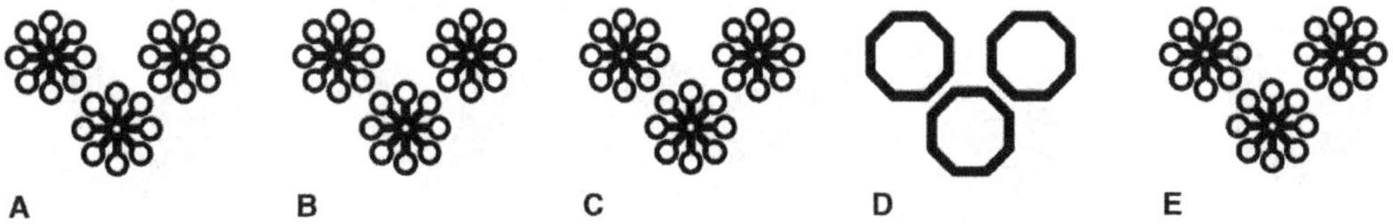

80.

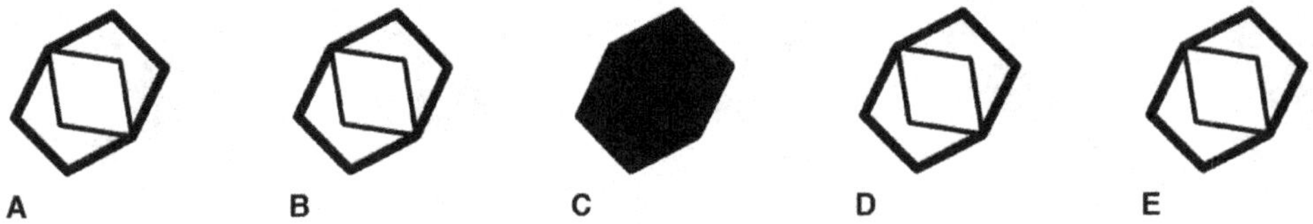

81.

A B C D E

82.

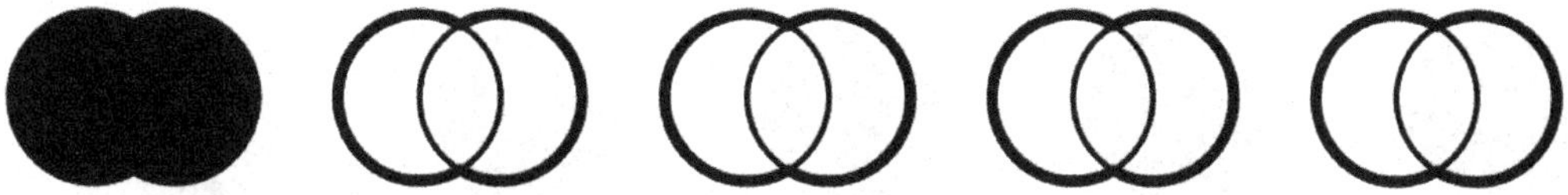

A B C D E

83.

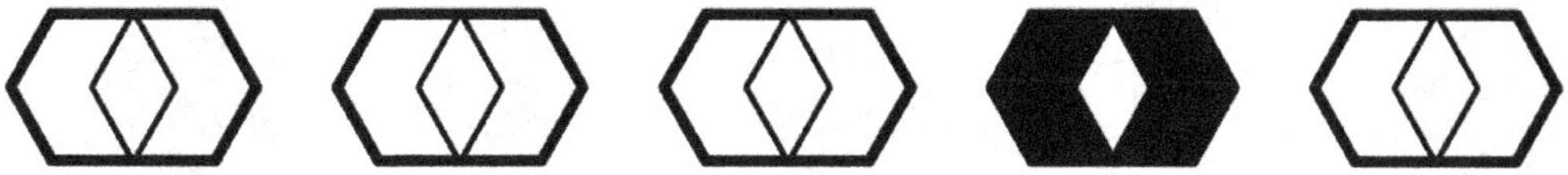

A B C D E

84.

A B C D E

85.

A B C D E

86.

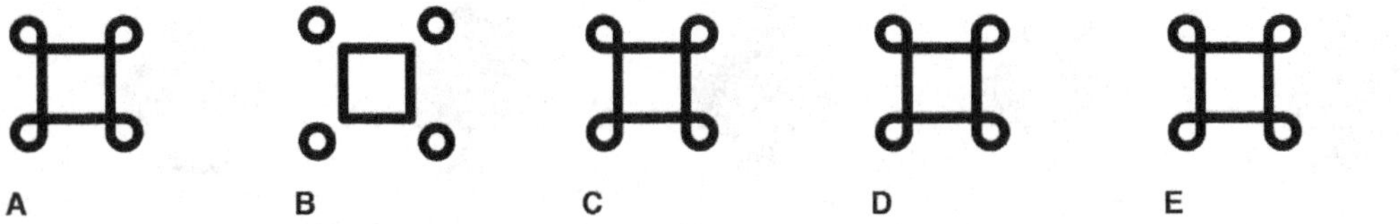

87.

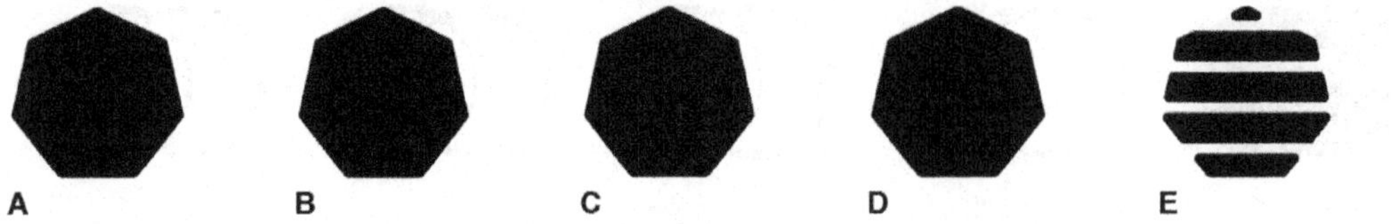

88.

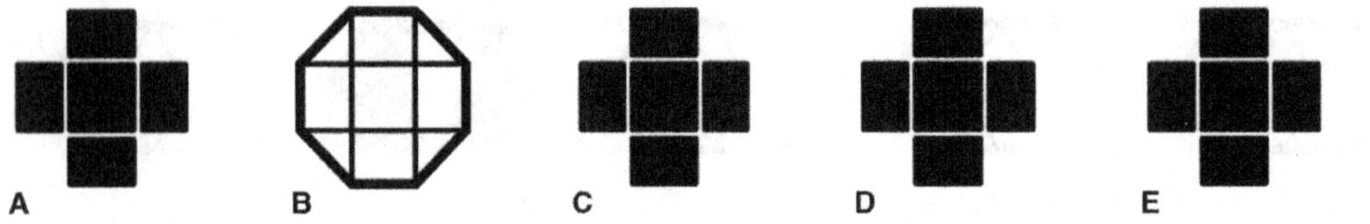

89.

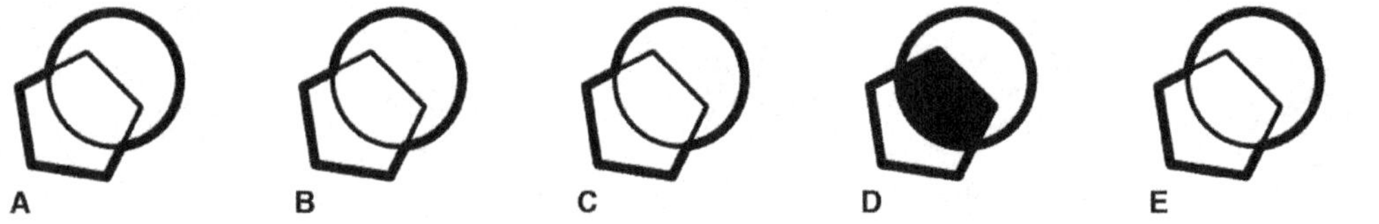

90.

91.

A B C D E

92.

A B C D E

93.

A B C D E

94.

A B C D E

95.

A B C D E

96.

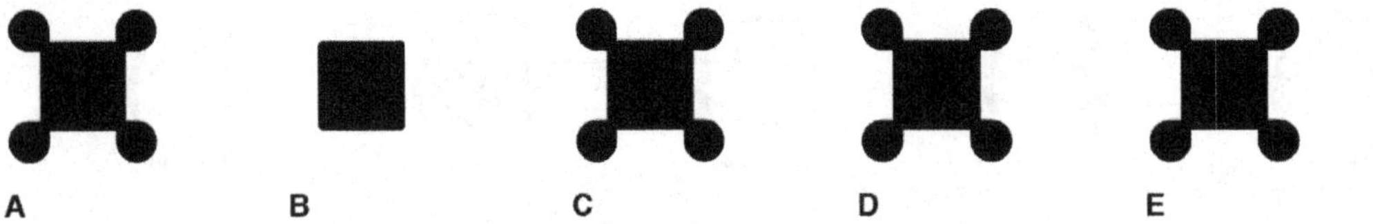

97.

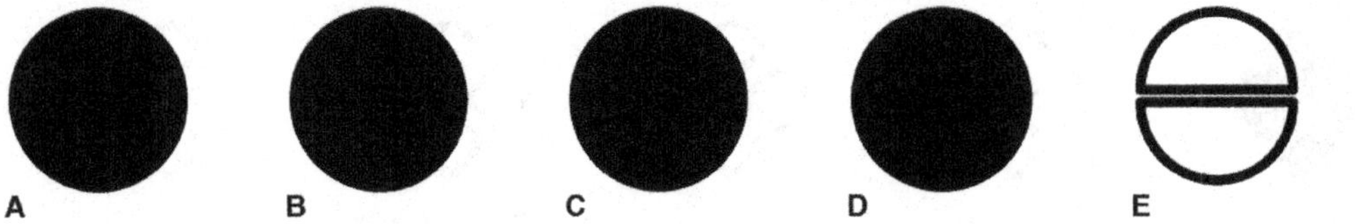

98.

99.

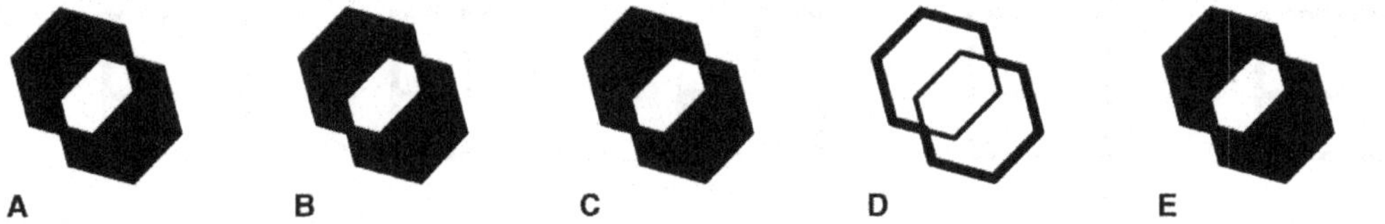

100.

Answers

1. D	26. A	51. D	76. B
2. B	27. D	52. B	77. D
3. D	28. B	53. E	78. A
4. A	29. A	54. E	79. D
5. D	30. D	55. B	80. C
6. B	31. D	56. D	81. E
7. E	32. B	57. A	82. A
8. B	33. D	58. E	83. D
9. B	34. D	59. B	84. B
10. D	35. A	60. D	85. C
11. B	36. D	61. D	86. B
12. B	37. A	62. A	87. E
13. A	38. D	63. D	88. B
14. C	39. B	64. A	89. D
15. E	40. E	65. B	90. A
16. E	41. A	66. D	91. D
17. A	42. D	67. C	92. A
18. B	43. B	68. B	93. D
19. D	44. E	69. A	94. C
20. A	45. B	70. E	95. B
21. C	46. B	71. E	96. B
22. B	47. D	72. B	97. E
23. A	48. A	73. D	98. B
24. D	49. B	74. A	99. D
25. E	50. D	75. D	100. A

Hidden Figures

Like the Simple Drawing section, the Hidden Figures section of the SIFT exam evaluates a test taker's ability to quickly and accurately assess visual information. However, the comparison ends there because the challenge posed by each section is quite different. In the Simple Drawings section, the clock and the pace are grueling. The solution to each question should be obvious and instead, the challenge lies in completing as many questions as possible in the allotted two minutes. In contrast, in the Hidden Figures section, the questions themselves are challenging but the pace is somewhat more forgiving. Test takers are given five minutes to complete 50 questions, each of which entails scanning a somewhat complicated drawing for a smaller embedded hidden figure or shape.

Although the time to complete each question is more liberal than in the Simple Drawings section (6 seconds per question versus only 1.2 seconds per question), most test takers still struggle to complete all of the questions in the allotted time, due to the increased difficulty of the questions. Whereas the solutions in the Simple Drawings section can be ascertained with a quick glance, those in the Hidden Figures section take more focus and consideration.

The Hidden Figures questions measure where each candidate falls on the Field Independence–Field Dependence Scale. Candidates whose score fall on the high end of the Field Independence end of the scale are able to focus on only the important information in their visual field, while ignoring the irrelevant information. Those whose scores fall closer to the Field Dependence end of the scale struggle to hone in on the relevant information and cues in the visual field and get distracted by unimportant information within the field of view.

Army admissions officers are looking for candidates that have a high degree of Field Independence because it is crucial that Army pilots can accurately and nearly instantly interpret visual information. There will certainly be times when Army pilots must fly in difficult conditions to drop cargo or pick up troop members in precise locations in a dense jungle or snowy mountaintop. In such situations, the ability to quickly scan one's visual field and hone in on important—and perhaps hidden—visual information while ignoring everything that doesn't matter is crucial. Achieving high scores on the various visual aptitude sections of the SIFT exam will demonstrate to Army admissions officers that a candidate is strong in these critical skills.

The Hidden Figures section of the SIFT contains 50 questions: ten sets of five questions. At the beginning of each of the sets, there are five different figures labeled A-E. Each of the 50 questions depicts a complex arrangement of shapes, patterns, and lines, which actually have one of the five figures (A-E) at the top of the set embedded within it. Test takers must identify and select the one lettered figure that is hidden in the puzzle-like image. It is important to note that the embedded hidden images will always be of the same size and orientation as they were at the top of the section; each question contains exactly one hidden image within it. However, even though there are five labeled images and five questions per set, not all images are necessarily hidden in one of the five questions. Likewise, images may be hidden in more than one question. For example, figure B may be hidden in question 3 and 4, while figure D is not hidden in any of the five questions.

As mentioned, even though the allotted time to answer each Hidden Figures question is more forgiving than in the Simple Drawings section, it is often more difficult to complete all of the questions in this section because the figures may be well-hidden and require some time to evaluate and locate. Some arrangements and patterns in this section can pose a challenge even for candidates with a high degree of Field Independence.

Unlike the Simple Drawings section, which relies almost entirely on speed, there are a couple of different strategies that can be used on the Hidden Figures questions. Certain strategies may be more appropriate and helpful for certain figures and puzzle arrangements as well as the needs and strengths of each test taker. Test takers should experiment with a variety of techniques to see what approach is most helpful for this section and should bear in mind that some combination of the two may be optimal.

Strategy 1: Some test takers start by examining the five labeled figures at the top of each set, and look for those with distinguishing features such as long straight sides or sharp angles or protrusions. From there, the complex arrangements of each question are evaluated, with a focus on uncovering the labeled figures that had a prominent feature by searching for the prominence. If the arrangement clearly cannot accommodate the prominent feature, that figure can be eliminated as a possibility and the remaining figures can be tried.

Strategy 2: Some test takers prefer to begin by systematically examining each of the puzzle-like arrangements in the set of questions. This strategy also tends to be better when the hidden labeled figures do not have very pronounced features from which they can be identified. While focusing on one arrangement (one question) at a time, the test taker tries to fit each of the labeled figures from the top of the set into the question one by one. Because there is only one figure hidden in each arrangement, as soon as a fit is identified, the answer is marked and they can move on to the next question. Again, it is important to note that figures may appear in more than one arrangement (so test takers should not cross the figure off after it is found in one question) and the figures, when hidden, will be exactly as they appear at the top of the set (same orientation, size, shape, etc.)

Practice Questions

A ___

B ___

C ___

D ___

E ___

1.

2.

3.

4.

5.

A ____

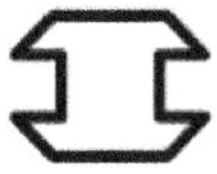

B ____

C ____

D ____

E ____

6.

7.

8.

9.

10.

A ___

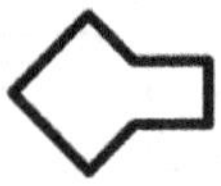

B ___

C ___

D ___

E ___

11.

12.

13.

14.

15.

A ___

B ___

C ___

D ___

E ___

16.

17.

18.

19.

20.

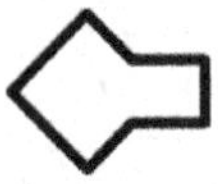

A ___

B ___

C ___

D ___

E ___

21.

22.

23.

24.

25.

A ___

B ___

C ___

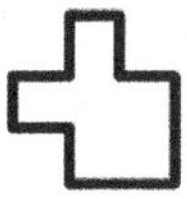

D ___

E ___

26.

27.

28.

29.

30.

A ____

B ____

C ____

D ____

E ____

31.

32.

33.

34.

35.

A ___

B ___

C ___

D ___

E ___

36.

37.

38.

39.

40.

A

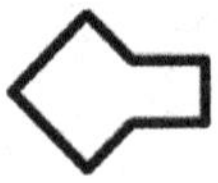

B

C

D

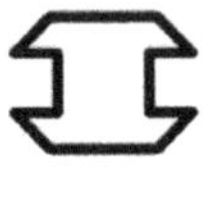

E

41.

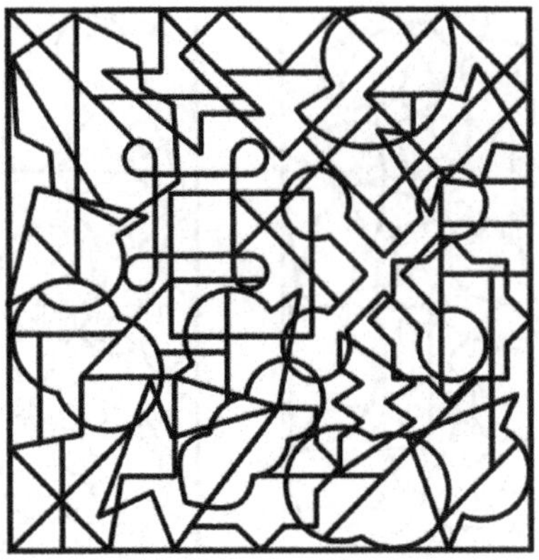

42.

43.

44.

45.

A ____

B ____

C ____

D ____

E ____

46.

47.

48.

49.

50.

Answer Explanations

A ___

B ___

C ___

D ___

E ___

1.

2.

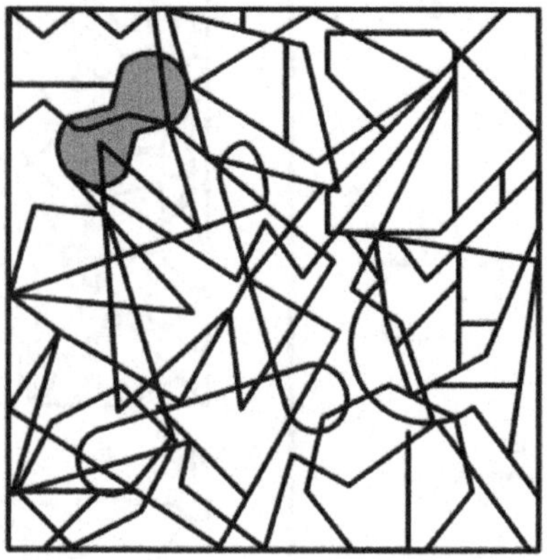

3.

4.

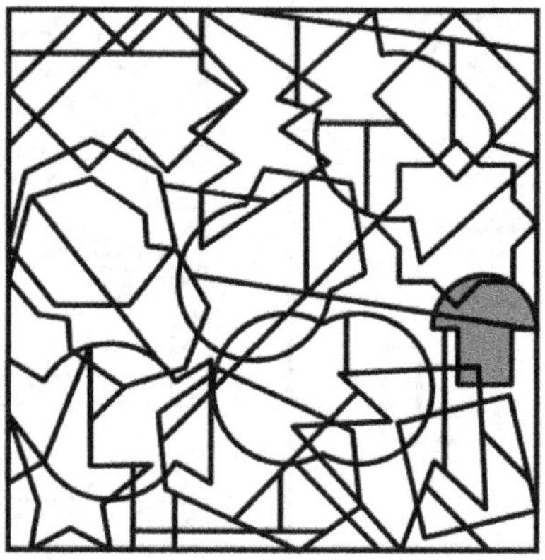

5.

A ____

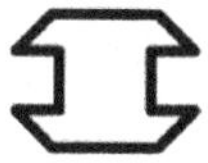

B ____

C ____

D ____

E ____

6.

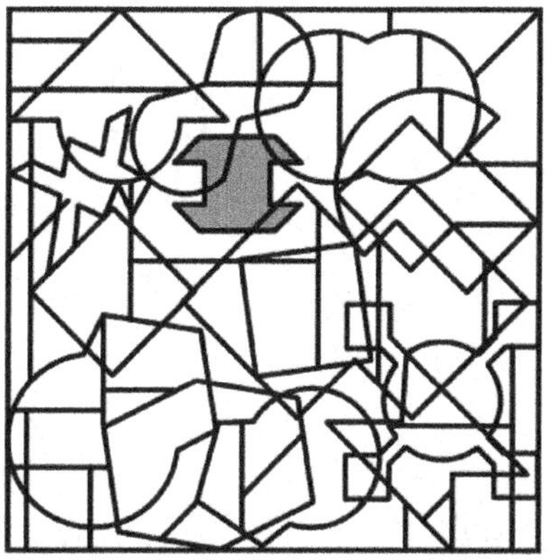

7.

8.

9.

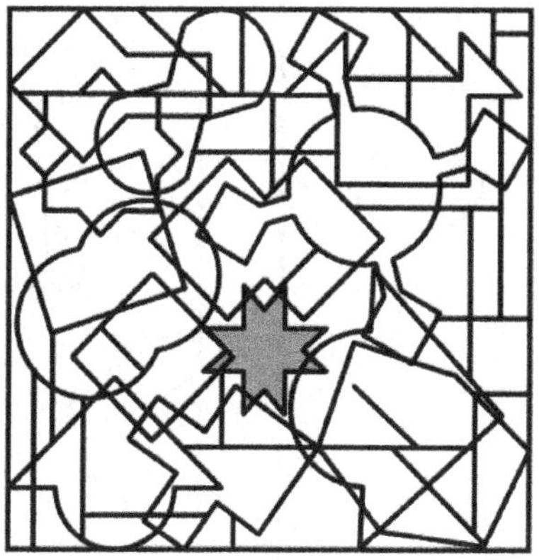

10.

A ___

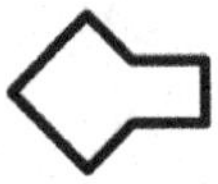

B ___

C ___

D ___

E ___

11.

12.

13.

14.

15.

A ___

B ___

C ___

D ___

E ___

16.

17.

18.

19.

20.

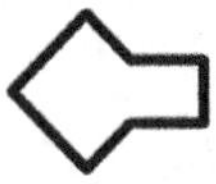

A ___

B ___

C ___

D ___

E ___

21.

22.

23.

24.

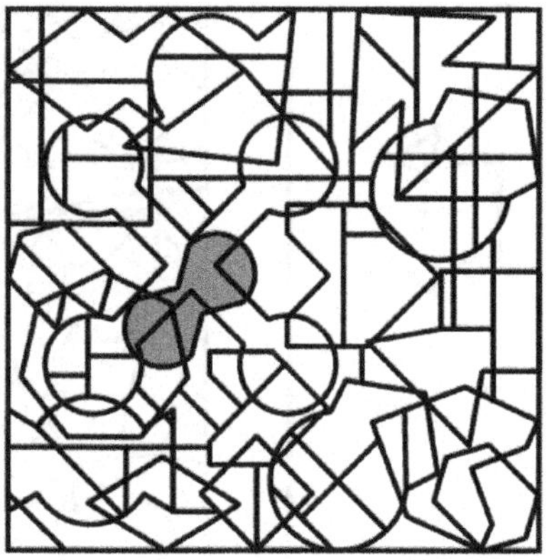

25.

A ___

B ___

C ___

D ___

E ___

26.

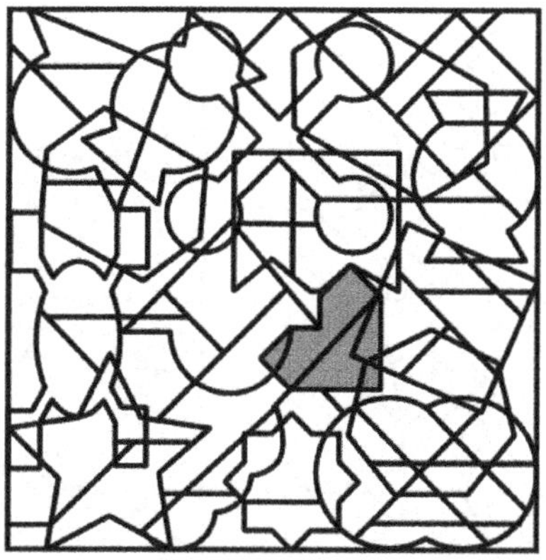

27.

28.

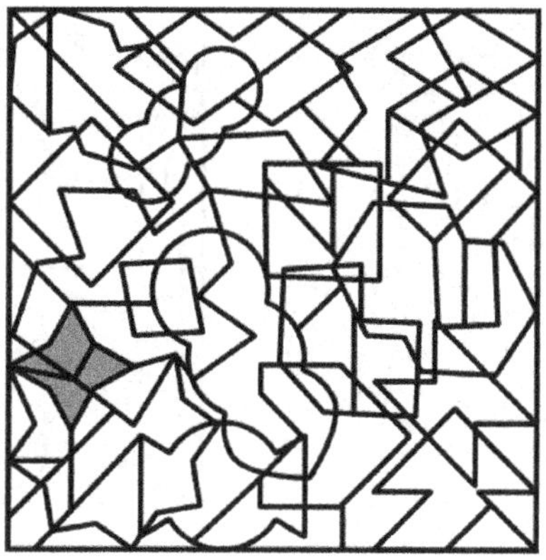

29.

30.

A ___

B ___

C ___

D ___

E ___

31.

32.

33.

34.

35.

A ___

B ___

C ___

D ___

E ___

36.

37.

38.

39.

40.

A ___

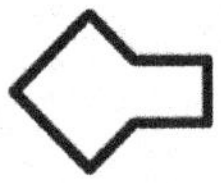

B ___

C ___

D ___

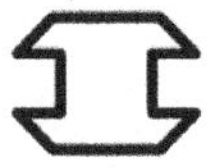

E ___

41.

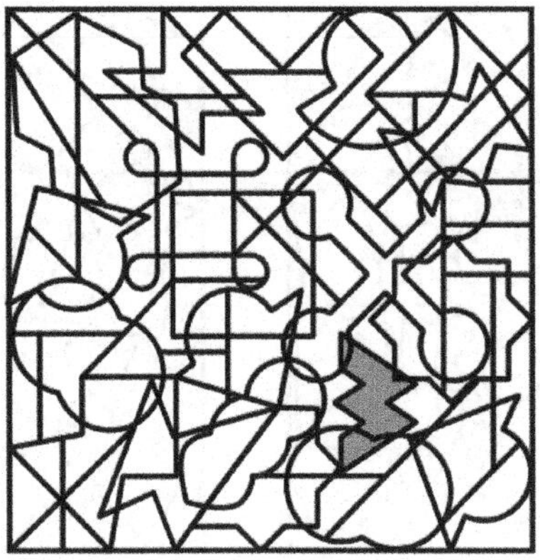

42.

43.

44.

45.

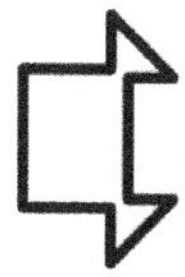

A ____

B ____

C ____

D ____

E ____

46.

47.

48.

49.

50.

Army Aviation

Aerodynamics

The Nature of Flight

Flight is not a natural state. All objects are affected by *gravity*, which constantly pulls things downwards—unless something else is providing support. A person sitting in a chair is certainly off the ground and, despite the constant pull of gravity, as long as he or she is supported by the chair, that's not likely to change. Even as gravity pulls down on the person, he or she collides with the chair, which then pushes the person up. It may seem rather confusing that an inanimate object has the ability to push up, but that is exactly what the chair does. If it didn't, then the chair would simply collapse under the person's weight, and the person would fall to the ground. Instead, the chair's rigidity keeps the chair intact, and the chair pushes upwards with enough force to maintain its shape and form. The ultimate takeaway message, here, is that in order to overcome the downward force of gravity, there needs to be some source of upward force to cancel it out.

Overcoming the force of gravity presents a unique challenge for an aircraft, as there isn't any chair equivalent that can conveniently lift a helicopter off the ground. Even if there were, this hypothetical object would also be affected by gravity and would, in turn, need to be supported by the ground. Obviously, this force has to come from somewhere. In the case of a helicopter, the force comes in the form of lift generated by an airfoil.

Lift

An *airfoil* (used in virtually all aspects of flight) is a blade; it cuts through the air so that air passes above and below it. Moreover, the air is not treated equally by the airfoil's design. The top of the airfoil is *cambered*, or curved, allowing the air to quickly get past the airfoil and create a low pressure. The air that goes below the airfoil has a shorter distance to travel and is slower, which results in high pressure. *Lift* is the upward force created by the difference in air pressure above and below the blade. It is what allows a helicopter to fly. A good way to visualize why this works is to consider a block between two springs:

When the block is directly between the two springs, the block doesn't move because the pressure from the springs is equal and opposite, canceling each other out. However, if one spring were to be compressed while the other was stretched, the compressed spring would suddenly be exerting more force, while the stretched spring would exert less. The block will be pushed until the springs are once equal and opposite. Likewise, the reduced pressure above the airfoil pushes *downward* with less force than the increased pressure below pushes *upward*, and lift is created.

The amount of lift created depends on several things, from the material and shape of the airfoil to the density of the air in the vicinity. A pilot can directly influence only two—the angle of incidence (pitch angle) and the angle of attack (AOA):

- *The pitch angle* is the angle between the *chord line* (a line drawn between the leading edge and the trailing edge of the airfoil) and the direction of the rotor's spin. The angle can be directly controlled using the collective on the helicopter (to be discussed in further detail, below).
- The *AOA* is the angle between the chord line and the direction of the incoming air. On a windless day, AOA and pitch angle can potentially be the same when hovering as the only relative wind will be that created by the rotation, but knowing the difference is important.

Typically, in order to create lift, a *positive AOA* is needed—the *front* of the blade needs to be raised above the *back*, relative to the wind. If the angle is *zero* (when the two are actually at the same height, relative to the wind), no lift will be generated. *Negative lift* is technically possible by flying at a *negative AOA*. There are exceptions; however, some airfoils (specifically those that are either *cambered* or *nonsymmetrical*) can be designed to generate lift with a zero, or even negative, AOA, though increasing the AOA will still result in more lift.

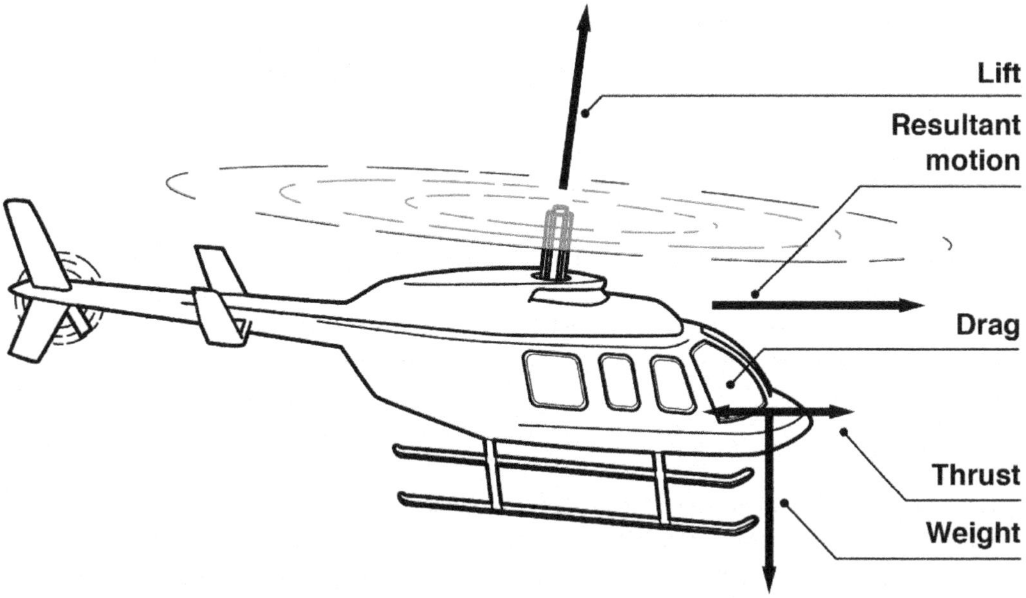

Weight

It's not enough to simply create lift. *Weight* is the force acting on a body due to gravity, and *gravity* is the acceleration acting on a body due to the Earth's mass. The average acceleration upwards must be larger than the average gravity acting on the body, and the total generated lift needs be a larger force than the loaded weight. The *weight of the aircraft* refers to the weight of the helicopter itself, all of its passengers, cargo, fuel, and anything else that would fall to the ground, if they were not supported by the helicopter. The heavier the aircraft, the greater the force needed to move it, thus requiring more lift to get off the ground. Of course, gravity does not stop because the helicopter is off the ground, and the helicopter must continue maintaining at least as much lift or it risks losing altitude.

There are several factors that can potentially increase the weight and ultimately increase the amount of lift needed to stay aloft. To say that the aircraft's *weight* has increased, however, is a bit of a misnomer as the mass of the helicopter does not generally change significantly midflight (other than the burning of fuel). Rather, some factors can either decrease the effectiveness of lift or increase the potency of weight. Either change requires that additional lift is created to compensate as if its weight had actually increased. Strong winds and other adverse weather conditions can cause this, but a more frequently observable example is the simple act of banking the helicopter.

When *banking* the helicopter, the lift is no longer generated directly downwards; rather, some of the vertical lift is redirected into horizontal movement. This results in a descent, unless the total airfoil is increased to produce the appropriate amount of vertical lift once more. While, at small angles, the effect is quite tame, the effect gets exponentially stronger for each increase in banking angle. For example, a bank of 25° would result in a 10% increase in required lift, but increasing the angle by just nine more degrees, to an angle of 34°, the helicopter will have already reached a 20% increase. By the time an angle of 48° is reached, the lift will have increased by half again of what an upright helicopter requires. Were a helicopter to bank all the way to 90°, the airfoils would be generating only horizontal lift, with no vertical lift, and so the helicopter would essentially be free falling.

The added strain created by this effect is often called the *load factor* (or the *G load*) because the strain increases the load, or weight, on the airfoil. The value of the load factor does not give an exact value for this additional weight, but rather is expressed as a percentage of the helicopter's resting weight. This value can be calculated using the following equation:

$$G = \frac{L_1}{L_0}$$

Here, G is the load factor, L_1 is the actual load on the rotor blades, and L_0 is the resting, or normal, load. So, a 1-ton helicopter with a load factor of 3 experiences the same force as a 3-ton helicopter hovering at rest. If the helicopter is not able to generate sufficient lift to offset this increase, the helicopter will begin to lose altitude unless the pilot can adjust for the weight in other ways, such as reducing the banking angle. For that reason, knowing how much lift the helicopter can produce is crucial. Otherwise, the helicopter may become overloaded, which will cause the rotor to droop, and the aircraft will descend.

Thrust

Lift and weight are perhaps the most important forces involved in the helicopter's movement, but if they were the only two, the helicopter would be little more than a glorified elevator capable only of rising and falling. Just as force is needed to lift the helicopter vertically from the ground, force is also needed to move the helicopter horizontally. This force, called *thrust*, can be in any direction, but is almost always measured in horizontal motion. Thrust is primarily created by converting lift via alterations of the pitch angle and AOA. Since horizontal movement does not have to contend with gravity, less effort is required to shift the helicopter horizontally compared to that required vertically. This means that only a small portion of lift needs to be sacrificed to achieve some movement, although more can be sacrificed if faster horizontal movement is desired.

Drag

Just as lift contends with the weight of gravity, thrust also has an opposing force to overcome, called drag. *Drag* is actually a composite force, usually broken up into three subcategories known as profile, parasitic, and induced. All three forces change in varying amounts with the speed of the helicopter and, in most cases, with increasing, comes increasing drag. The faster the helicopter travels, the more thrust force is required to overcome the increased drag. However, there is a narrow but important exception, which will be discussed in more detail below.

Anyone who has ever ridden a bike or stood outside on a windy day has experienced drag. Drag is simply the sensation of the wind exerting force on objects that the wind passes through. The strength of that drag depends only on the relative speed of the wind, such that standing still in 20 MPH winds creates just as much drag as moving 20 MPH in still winds, or even moving 10 MPH into a 10 MPH wind. Regardless of the exact combination, the result is the same—both the object and the air attempt to occupy the same space, and the object must exert force to displace the air, even as the air tries to do the same to the object. The faster the wind is moving relative to the object, the more air that object has to push out of the way and the more air that pushes back on the object. This explains why increasing velocity also increases drag.

- *Parasitic drag* is the simplest of the three subtypes of drag and functions exactly as explained above. It is caused by the non-lifting portions of the helicopter, such as the fuselage, engine cowlings, hub, mast, landing gear, and external loads. Specifically, parasitic drag increases with airspeed and is the dominant type of drag at high airspeeds.

- *Profile drag* is similar to parasitic drag, but with one critical difference—it only concerns the drag created by the frictional resistance of spinning airfoils. Since the airfoils are much smaller than the rest of the helicopter and designed to be as aerodynamic as possible, profile drag is a lot smaller than the high speeds of the rotor blades might suggest, although the drag is still significant at lower speeds. Drag does increase as velocity increases, but only by a small margin, because relative to the speed at which the rotor blades are spinning, the velocity added by a moving helicopter only represents a small percent increase to its effective speed. Going from 5 MPH to 10 MPH will quadruple parasitic drag, but the rotor's speed is changing from approximately 605 MPH to 610 MPH, which would only increase the rotor's drag by less than one percent. This is not entirely accurate; since the airfoil is rotating, the actual speed of the rotor blade relative to the air changes, based on how far down the blade contact occurs and what direction the blade is currently spinning. However, this example should at least explain why profile drag increases so slowly relative to parasitic drag. When compared to the other two types of drag, profile drag remains midway throughout—neither dominating nor negligible.

- *Induced drag* is a byproduct of the lift created by the rotor, and its very existence is unique. As the blades spin, they briefly leave a small gap of empty space behind them where they once were, which is quickly filled by the air above and below the blade. Since the air is at a different pressure above and below, the air creates a spiral directly behind the blade, which pushes air in the opposite direction of lift. When the lift the helicopter creates is at an angle (such as is the case while banking), the induced drag is too, which causes some of the normally-downward directed force to instead face rearwards. This rearwards force pushes the helicopter back as the helicopter tries to move forward. The greater the AOA is, the greater the portion of the force that is directed rearwards. Since the angle of attack is usually greatest at lower airspeeds and reduced at higher airspeeds, this is the only drag that typically decreases as airspeed increases. When compared to the other two types of drag, induced drag is the primary force at low speeds, but at higher speeds becomes barely meaningful.

The *total drag* on the helicopter is simply the sum of the three subtypes. The graphic below shows how the individual components of drag add up at different speeds:

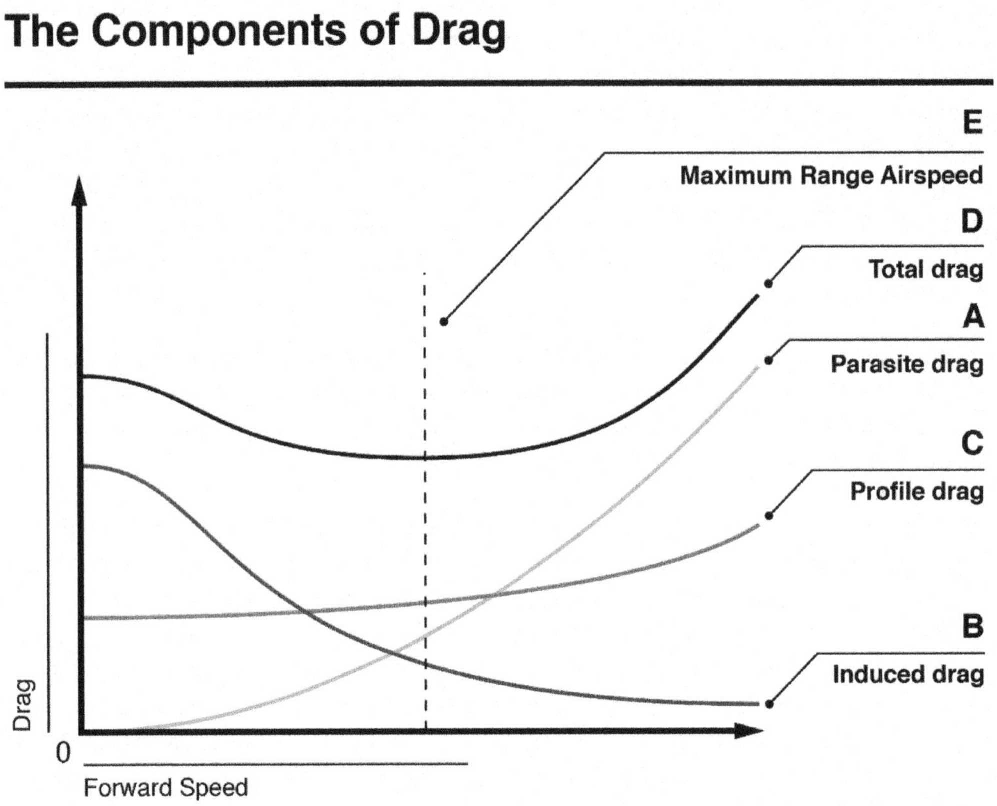

Induced drag generally contributes the most at lower speeds, while parasitic drag dominates at higher velocities. This unique behavior means that most helicopters have an optimal velocity, wherein induced drag has decreased significantly, but parasitic drag has not yet taken over. This speed is important because this value is the point at which the lift to drag ratio is highest, and is often used to determine a helicopter's best rate of climb and other numerical figures. This point is called the *maximum lift/drag ratio*, or $\frac{L}{D_{max}}$. Point E on the drag/airspeed relationship diagram represents the maximum range airspeed. This airspeed will allow the helicopter to fly the furthest distance on a tank of gas.

Weight and Balance

In discussing the forces acting upon the helicopter, it's important to expand upon weight—perhaps the most important force to consider when flying—and how weight relates to another concept: balance.

Weight

The basic concept of weight has already been discussed earlier as it exists principally as the opposing force to lift. This section will focus on the more technical aspects of weight and how weight pertains to helicopters. The heavier something is, the more difficult that object is to move, thus the greater the lift required to get off the ground. Even if the helicopter is theoretically capable of generating the lift needed to do so, the strain of an overloaded helicopter can cause structural damage to the helicopter.

When talking about helicopters, there are several terms that are often used when discussing weight: basic empty weight, maximum gross weight, and weight limitations.

- *Basic empty weight*, as the name suggests, is the weight of the helicopter once everything directly unattached to the helicopter is removed. More technically, the basic empty weight solely includes the weight of the helicopter itself, any optional equipment, and all the fluids contained within the components themselves, including unusable fuel, transmission oil, hydraulic fluid, etc. As a general rule of thumb, if the helicopter could still function properly without something, that something isn't included in the basic empty weight.

  Some helicopters may also have a different value listed called the *licensed empty weight*, which is the basic empty weight with the engine and transmission oil excluded. If this is the case, the oil's weight will need to be accounted for and added manually when determining the actual weight of the helicopter.

- The *maximum gross weight* is the maximum weight that a helicopter can bear safely. This weight is often broken up into an internal maximum gross weight and external maximum gross weight.

  - *Internal maximum gross weight* is the maximum weight within the interior of the helicopter, including the pilot, passengers, and any gear or baggage.

  - *External maximum gross weight* is the weight that can be supported externally by the helicopter, including external winches. When properly balanced, the external maximum gross weight can reach tremendous levels.

- It is very important for a pilot to understand the *weight limitations* on the helicopter. As weight increases, the power required to produce the lift needed to compensate for the additional weight must also increase. By reducing weight, the helicopter is able to safely takeoff or land at a location that would not be possible for heavier aircraft. If sufficient takeoff power is questionable, the takeoff should be delayed until the aircraft is lighter or the density altitude has decreased. If airborne, a landing zone that favors better conditions, or that does not require landing to a hover, would be a safer choice. Also, aircraft operating at higher gross weights are required to produce more torque. The additional torque requires more tail rotor thrust to compensate for the main rotor torque effect. Some helicopters may experience loss of tail rotor thrust during high altitude operations.

  If the helicopter is packed too lightly, it can easily become unbalanced with the sole addition of the pilot, which can, in turn, make controlling the helicopter more difficult. This is easily remediated by adding extra weight in the rear of the helicopter to act as ballast. The goal is ultimately to balance the weight, using the center of the rotor disk (for single rotor helicopters) or some space between them (for multi-rotor helicopters). This need for balance connects to the concept of the center of gravity.

Center of Gravity

The *center of gravity* is the point at which the entire weight of the helicopter, including both internal and external attachments, is averaged. It's helpful to visualize an unweighted seesaw to understand how this works.

With no other forces acting on it, the seesaw will be perfectly balanced, and the center of gravity is on the pivot point or the fulcrum. When a ten-pound weight is added to one side, there will be more weight on one side of the pivot; the center of gravity is no longer on the pivot, and the seesaw begins to tip. Putting a second, identical weight opposite of the first would cancel the two out. The weight can also be canceled with a larger weight placed closer to the fulcrum or a smaller weight placed further away.

Seesaw with weight

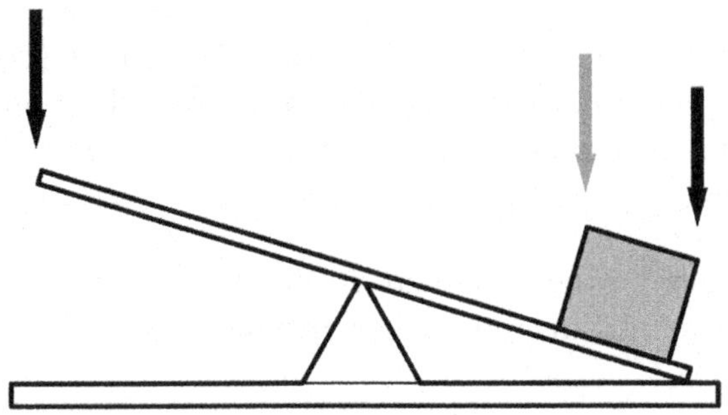

A helicopter acts similarly to a seesaw, with the rotor disk acting as the pivot point. Too much weight on one side will cause the helicopter to tip. Unlike a seesaw, however, a helicopter can adjust the angle of its rotor disk in order to stabilize, but doing so comes at the cost of some lift. This lost lift, in turn, results in reduced performance, and reduced performance makes every other task that much more difficult. Therefore, knowing how to spot and solve center of gravity problems is crucial.

While lateral center of balance problems can occur, they are less frequently a problem when compared to longitudinal center of gravity issues. This is primarily due to the shape of helicopters, which are generally longer than they are wide, so far more weight is required to disrupt the helicopter's stable flight. Laterally-imbalanced centers of gravity are occasionally created as some helicopters might dictate which seat to use when flying solo. When using a side mounted winch, or something similar, the weight can quickly place the helicopter outside of the center of gravity limits. However, a front-loaded or aft-loaded center of gravity is much more common.

- A *front-loaded center of gravity* can be identified by the fuselage tipping forward, with the nose pointing down towards the ground. This will cause the helicopter to pull forward when hovering and makes slowing down from forward movement more difficult.

  This is most often the result of having a heavy pilot at the front, with little-to-no cargo in the rear. A front-loaded center of gravity can also occur midflight, if the fuel tank is located in the rear. As the fuel is consumed, the weight in the rear decreases, and the front begins to tip.

  A common solution to a front-loaded center of gravity is to add weight to the back of the helicopter, until the fuselage is once again level. If a longer flight is expected, it may be beneficial to add weight to the rear of the helicopter in anticipation of the fuel tank decreasing in weight.

- An *aft-loaded center of gravity* can be identified by the fuselage tipping backwards, with the nose pointing up towards the sky. This will cause the helicopter to pull backwards when in a neutral hover position and will make accelerating forward more difficult.

An aft-loaded center of gravity, in conjunction with strong winds, can even potentially tilt the helicopter sufficiently that the tail bloom is forced into the rotor disk when attempting to move forwards. This is a fringe case, requiring both a significantly aft-loaded center of gravity and a compromising factor, such as strong winds or poor technique. Pilots should be cognizant of the risk of this condition, even if two of the three conditions are met. An aft-loaded center of gravity is most often the result of having a light pilot with a large amount of cargo in the rear, possibly including a full fuel reserve.

A common solution to an aft-loaded center of gravity is to remove weight from the back of the helicopter and add weight to the front until the fuselage is once again level. Some helicopters have a forward tilt built into the mast to compensate for an aft-loaded center of gravity.

Recognizing a front- or aft-loaded center of gravity is much easier when doing so outside the craft than doing so while within the helicopter. For this reason, using ground personnel to observe the helicopter to spot a displaced center of gravity can be very beneficial, but if that's not possible, the signs of poor center of balance can still be observed from within. If there are no strong winds, a front- or aft-loaded center of gravity can be identified by the helicopter's tendency to pull forward or backwards, respectively. If strong winds are present, separating idle movements caused by the winds and those caused by the center of gravity can be exceedingly difficult. If that is the case, the horizon will need to be used as a level guide as the helicopter lifts off.

Flight Controls

Now that the principles of flight have been covered, the next step is to understand the controls of flight. There are four primary controls that are used to control the helicopter: the collective pitch control, the throttle, the cyclic pitch control, and the anti-torque pedals. Each has a specific function in moving the helicopter through the air, except for the throttle, whose purpose is to control the engine's power output.

Flight controls

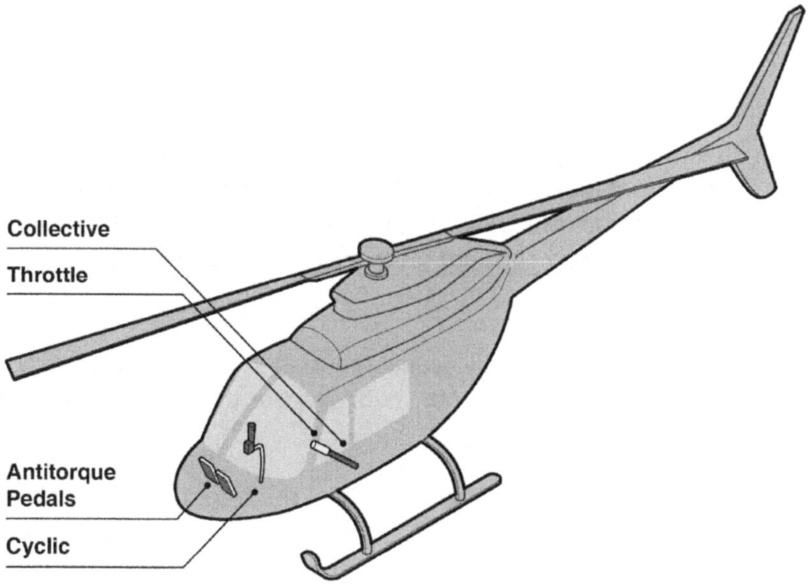

The Collective Pitch Control

The *collective pitch control*, often simply called the *collective*, is primarily used to create lift and control vertical movement. The collective is typically located to the left of the pilot's seat and can be raised or lowered to increase or decrease lift, respectively.

Mechanically, the collective achieves this by altering the pitch of the rotor blades simultaneously (collectively). By increasing the pitch of the blades, the angle of incidence is also increased, and more lift is generated. Conversely, decreasing the pitch of the blades results in a lower angle of incidence and less lift. However, increasing or decreasing lift will also have an identical effect on the drag on the blades. This drag will slow down the rotations per minute (RPM) of the helicopter, unless compensated for by increasing the power to the engine using the throttle.

The Throttle

The *throttle* is used to control the power to the engine. Rotor RPM needs to stay within a certain range to operate the helicopter safely; the throttle is used whenever the RPM drops or rises beyond that value. Using the throttle is much like a motorcycle throttle, with a clockwise turn decreasing power and a counterclockwise turn increasing power.

Since sudden and drastic changes can quickly lead to a destabilized flight, slower and smoother adjustments to the throttle are ideal for maintaining control of the helicopter. Making large, sudden adjustments often leads to over-controlling, where the desired position is passed, and an opposite adjustment is needed to correct the mistake. This is actually excellent advice for all the controls, but is particularly true for the throttle, because the throttle controls the all-important RPM.

Although every helicopter has a throttle, not every helicopter requires manual use of the throttle. Some helicopters are equipped with either a correlator or a governor, which is a tool designed to simplify throttle control. Both are designed to maintain RPM so the pilot does not need to, although they do so in different ways.

- A *correlator* is a mechanical tool that links the collective to the throttle, automatically increasing and decreasing the throttle as the collective is used. While effective, the correlator does not account for any other factor that could decrease RPM, so some manual control is still occasionally required.
- The *governor* includes an electronic sensor that tracks the current RPM directly, rather than using the position of the collective to guess. If the governor detects that the RPM is out of the desired range, it will automatically adjust the throttle as needed. Since the governor measures the RPM directly, it can account for any variable that might impact RPM. In a properly set-up governor system, the pilot rarely needs to touch the throttle at all.

The Cyclic Pitch Control

The *cyclic pitch control*, or the *cyclic*, is primarily used to generate thrust and control horizontal movement. The cyclic is usually located between the pilot's legs or the two pilot seats and can be pushed in any direction.

Like the collective, the cyclic functions by altering the pitch of the rotor blades. However, unlike the collective, the cyclic does not increase the pitch of all the blades at the same rate, but rather cyclically. In other words, the cyclic alters the pitch of the blade based on where that blade is in the disk. The cyclic might decrease the pitch of the blades as they approach the aft, only to increase the pitch again as the blade nears the front. This results in the lift being generated unequally, which, in turn, causes the disk to tilt. Since generated lift is always perpendicular to the rotor disk, tipping the disk in this way results in a small, horizontal component to the lift, which creates the thrust needed to move.

For example, if forward movement is desired, the pilot pushes the cyclic forward. In a counterclockwise-spinning rotor blade, this causes the blades' pitch to swivel as they rotate, keeping a low pitch as the blade moves over on the right and increasing to a higher pitch as the blade moves by the left. In a clockwise-spinning rotor blade, the pitch is reversed. This results in the rotor blade tipping forward, converting a small portion of the generated lift into forward thrust.

The Anti-Torque Pedals

The *anti-torque pedals* are located at the feet of the pilot, much like the gas and brake pedals of a car. The purpose of this flight control mechanism is to control the pitch of the tail rotor blades.

As the main rotor spins, the rotor creates torque that spins the helicopter in the opposite direction to counteract. The tail rotor does this by creating thrust at the tail of the helicopter. Since this thrust pushes so far away from the center of gravity, rather than simply pushing the helicopter to the side, the anti-torque pedals instead create torque in the opposite direction than that of the main rotor.

The anti-torque pedals themselves allow the pilot to alter the pitch of the tail rotor, which, in turn, increases or decreases the torque the rotor creates. This has two purposes.

1. The first purpose is to counteract the torque of the main rotor. As the main rotor's pitch increases or decreases, its torque does as well, so the tail rotor must also be able to increase or decrease its opposing torque in order to keep the helicopter's heading constant.

2. The second purpose is by willfully allowing the tail rotor's torque to increase or decrease past the main rotor's torque, a pilot can create a controlled rotation that allows the helicopter to change to a new heading.

The anti-torque pedals are designed so that pushing the left pedal creates more torque in a counterclockwise direction, and pushing the right pedal creates torque in a clockwise direction. In a counterclockwise rotor helicopter, this means that the left pedal increases the pitch of the tail rotor, while the right pedal decreases the pitch. This is because the torque generated by the main rotor is clockwise, so the tail rotor must generate counterclockwise torque to counter the main rotor, and the pitch determines the strength of that counterclockwise torque.

It may help to remember that no net torque is not the same as no torque at all. Even when flying perfectly straight, there are two torques acting on the helicopter: one from the main rotor and one from the tail rotor.

This scenario is not that different from the example used when discussing lift—the block between two springs. The only difference is that now, rather than having one compressed spring and one stretched spring, there are two compressed springs, both attempting to push the block. The anti-torque pedals effectively control the compression of one of the springs. By reducing the compression, the other spring can push the block one way, and by increasing the compression, the controlled spring can overpower the other spring, moving the other way instead, like a game of reverse tug-of-war. Replace compression with torque, and one has the anti-torque pedals.

Some helicopters, rather than having a tail rotor, have multiple main rotors that spin in opposing directions. This allows the counter-rotating main rotor systems' torques to cancel each other out, removing the need for the tail rotor. Such helicopters do still have anti-torque pedals, but they alter the pitch of the main rotors, rather than altering the pitch of tail rotors. Specifically, the right anti-torque pedal increases the pitch of all counterclockwise rotor blades, which increases the clockwise torque, and the left pedal increases the torque of all the clockwise rotor blades, which increases the counterclockwise torque. In this way, the pedals can serve the same function, despite not having a tail rotor.

Basic Maneuvers

The Four Fundamentals

There are many different maneuvers, but all can ultimately be broken down into what are known as the four fundamentals of flight—straight-and-level flight, turns, climbs, and descents. Learning any maneuver is just a matter of learning the specific combination required by that maneuver.

- *Straight-and-level flight* is perhaps the simplest maneuver as once a helicopter has reached the desired speed, only minimal control is required to keep moving. Most of the pilot's effort will be directed towards keeping the helicopter level, which means using the various controls to make sure that the heading, airspeed, and altitude remain constant and that the helicopter's course relative to the ground remains in a straight line. An exception is when this fundamental maneuver is combined with another, such as a climb or descent. Even then, care should be taken to ensure that any adjustments that are made to the flight path are done slowly and smoothly.

- *Turns* are unique among the fundamentals as they have two methods depending on the speed of the helicopter. If the helicopter is flying at low speeds (or even just hovering), turns are performed using the anti-torque pedals, while at higher speeds, the cyclic is used to bank the helicopter. In either case, the goal of the turn is to change the helicopter's heading. A perfect turn would result in no altitude or airspeed change, and the rate of turning would remain constant, aside from the initial acceleration and final deceleration.

  The reason for this difference is due to the *laws of motion*, which state that an object in motion will remain in motion. The anti-torque pedals change only the heading of the helicopter, not the velocity, so while in motion, attempting to turn using only the anti-torque pedals will result in the helicopter flying in the same direction, only sideways.

  To actually change the direction of flight, the helicopter needs to change the direction of the generated thrust, which is where banking comes in. By tilting the rotor disk to the side, the helicopter creates sideways thrust, which accelerates the helicopter in the new desired direction.

- *Ascent* and *descent* are two sides of the same coin and follow a very similar procedure. During both, the collective will be used to alter the pitch of the main rotor blades, which, in turn, will change altitude either up or down. When the collective is increased in counter-clockwise rotating helicopters, the nose of the helicopter tends to pitch up, and the pilot is required to apply forward cyclic to maintain the same airspeed. In addition, when power is increased, the nose of the helicopter will turn right, due to the torque increase, and the pilot is required to increase left pedal input to maintain the desired heading and trim. When the collective is reduced, the opposite occurs. The nose pitches down and to the left. The pilot has to apply aft cyclic and the right pedal to compensate for these aerodynamic changes.

Hovering

Hovering is perhaps the most common technique employed by helicopter pilots, and the ability to do so is what distinguishes helicopters from many other aircraft. The goal of a hover is to position the helicopter so that the helicopter remains a set distance off the ground (typically just a couple of feet) and then remains stationary. This is, unfortunately, not as easy as it sounds, as a helicopter is anything but stable. The helicopter cannot simply be brought up to a desired height, then parked in the air like a car might be parked in a driveway. Instead, the controls must be constantly fined-tuned to keep the helicopter at a stationary hover.

The collective is usually the easiest to adjust, as the collective acts as a position control. A *position control* is any control that controls position, rather than velocity or acceleration. The handles on a foosball table are an excellent example of a position control, with the figure moving at a rate proportionate to the handles. Due to a phenomenon called *ground effect*, this is similar to how the collective works, in that raising the collective will cause the helicopter to rise a proportionate amount and then stop.

This also makes the collective less susceptible to over-control, as the collective must be raised significantly before the helicopter overcomes the ground effect and begins rising freely. Since the goal of hovering is to get to a certain altitude and then stop, this is ideal behavior, and once the desired altitude is reached, the collective is not likely to need much adjustment.

The anti-torque pedals, meanwhile, are a *rate control*, which means that the pedals control the rate of change. An example of a rate control is the gas pedal on a car, with the car accelerating faster the harder its pedal is pushed and the car slowing down when the pedal is released. Similarly, pushing the anti-torque pedal causes the helicopter to begin rotating in that direction at a rate determined by the pressure on the pedal. Releasing the pedal causes the rotation to slow to a stop.

It is ideal in a hover to use only minimal pressure on the pedals to keep the helicopter's bearing constant. Otherwise, there is the chance that the pedals might be over-controlled, with the helicopter unable to stop rotating before reaching the desire point. On calm days, the anti-torque pedals might not even be needed to maintain a hover, but on windy days, the thrust generated by the tail rotor can become inconsistent, requiring far more manipulation to steady.

By far the most difficult, however, is the cyclic. Whereas the collective is position-controlled and the anti-torque pedals are rate-controlled, the cyclic is *acceleration-controlled.* Manipulation of the collective results in movement to a particular position, while pressing the anti-torque pedals to continue to spin the helicopter until released. In comparison, use of the cyclic causes the helicopter to speed up as long as the cyclic is held, but upon release, the helicopter will continue to move in that direction at a similar velocity until the cyclic is pushed in the opposite direction to create a negative acceleration and decrease the velocity to zero.

For example, if a horizontal drift is noticed, the cyclic must be pushed gently in the desired direction of movement for a brief moment, released, and then reapplied in the opposite direction as the destination is reached. If the cyclic is only pushed once, the helicopter will continue to move in that direction until drag eventually slows the helicopter to a stop. This makes the cyclic doubly-easy to over-control as there are twice as many opportunities to do so, and acceleration is more difficult to perceive than position or speed. This also means that the helicopter is almost always moving slightly, and controlling the cyclic in such a way that the helicopter's velocity is truly zero is nearly impossible. Even if the helicopter did reach zero velocity, it would only take the slightest of breezes to shift the helicopter slightly and begin the process anew.

Vertical Takeoff to Hover

In a *vertical takeoff to hover*, the objective is to start from a landed position and end with the helicopter hovering just a few feet off the ground. Before beginning this maneuver, there are a few steps that must be done outside the helicopter.

The first task is to make sure that the helicopter has the necessary clearance to take off. The next step is to make sure that the area around the helicopter, particularly above and to the left and right if the aircraft, is free of any obstructions. Once that has been done, takeoff can proceed.

With the collective fully down, the cyclic should be placed in a neutral position, and the collective should be increased smoothly. Pedals should be applied to maintain heading, and the cyclic should be coordinated for a vertical ascent. As the aircraft leaves the ground, the proper control response and aircraft center of gravity (COG) should be checked.

Hovering Turn

In a *hovering turn*, the object is to change the heading of the helicopter while maintaining a constant position relative to the ground. Neither a horizontal nor a vertical position should shift during the maneuver. The anti-torque pedals are the primary controls needed for this maneuver.

To perform the hovering turn, light pressure should be applied on the anti-torque pedals in the direction desired. The helicopter should begin to rotate in that direction. Just like a hover, keeping the helicopter otherwise stationary is important, which means the cyclic and collective may need to be used to prevent the helicopter from drifting. However, doing so should be no more difficult than before.

When the helicopter nears the desired heading, the pedals should be released, and the helicopter's rate of turn should begin to slow and eventually stop. If the pedals have been used lightly, this should take very little time, and the pedals can be released moments before reaching the target heading without fear of overshooting the target heading. If the pedals are pushed with more pressure and the helicopter is allowed to rotate at a faster rate, considerably more space will be required. This is because the helicopter will not only turn further in a set amount of time, but also require more time to stop. Therefore, a slower, steady rate of turn is preferable.

The above all assume a relatively windless day, as strong winds can greatly impact a hovering turn. While turning away from the direction of the wind, the tail turns into the wind, which is more difficult due to the increased drag, so it requires more pressure to maintain turning speed.

When the helicopter reaches a parallel position, the tail tends to *weathervane* and may even be pulled with the wind. If this happens, it may be necessary to switch the pedals to keep the helicopter from turning too swiftly. This switching-over is possibly the hardest part about a hovering turn, as the sudden shift requires an equally swift shift in the pedals.

In addition to the difficulty from trying to find the proper pedal position to keep the helicopter from spinning too quickly, the sudden change in forces often upsets the balance of the other controls, requiring smaller, but still significant, adjustments simultaneously. The graphic below shows a theoretical 360-degree turn and the necessary adjustments to keep the helicopter moving consistently throughout every 90° increment.

The Necessary Adjustments for a 360° Turn

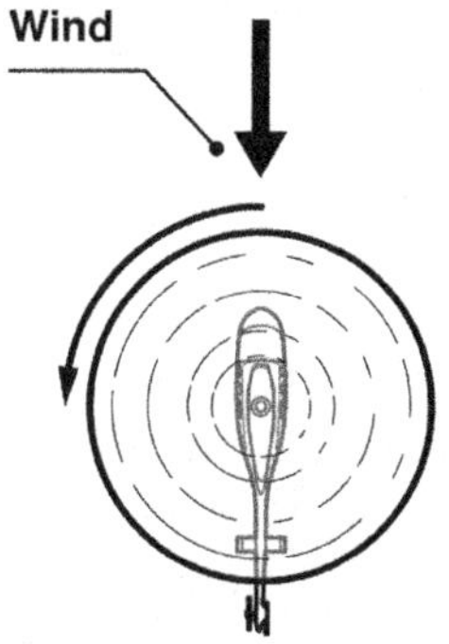

Cyclic - Forward

Pedal	Some left in hover, more left to start turn to left
Collective	Power required to hover at desired height
Throttle	As necessary to maintain r.p.m.

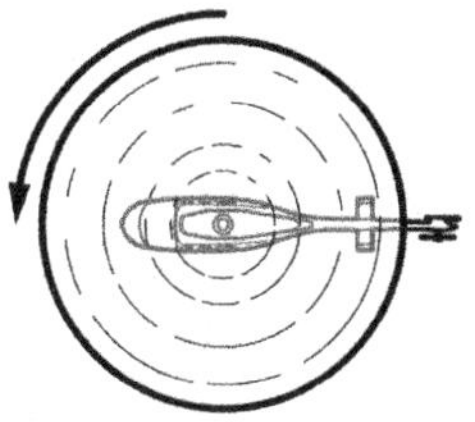

Cyclic - Right

Pedal	Most left pressure in turn
Collective	Most power in turn
Throttle	As necessary to maintain r.p.m.

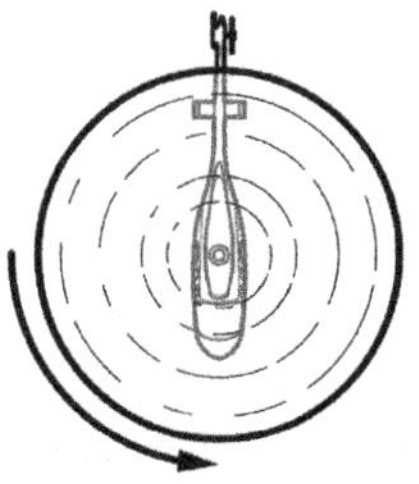

Cyclic - Rearward

Pedal	Changing from left to right pressure
Collective	Power reducing
Throttle	As necessary to maintain r.p.m.

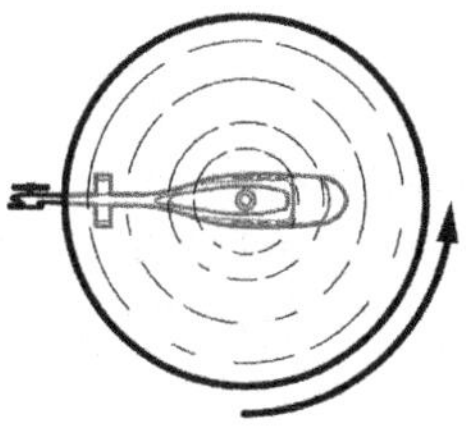

Cyclic - Left

Pedal	Most right pedal pressure in turn
Collective	Least power in turn
Throttle	As necessary to maintain r.p.m.

Cyclic - Forward

Pedal	Some right to stop turn, then left to maintain heading
Collective	Increasing as left pedal applied
Throttle	As necessary to maintain r.p.m.

Hovering to Forward Flight

In a *hovering to forward flight* maneuver, the helicopter will be leaving a hovering position to move forward. This maneuver is not designed to be a quick flight maneuver and is not to be used for long travel distances. Instead, the purpose of this maneuver is to reposition the helicopter or to travel short distances. Ideally, the helicopter should not exceed the speed of a brisk walk during its execution.

Starting from a hovering position, the first step is to make sure that the intended path is cleared and that helicopter is properly lined up with the desired target. If not, the pilot can perform a hovering turn to align the helicopter correctly. Once that is done, two points directly between the helicopter and the destination should be mentally identified. These are the *reference points* and will be used to ensure the helicopter does not drift during the maneuver.

To begin, the pilot should apply slight pressure to the cyclic. As previously mentioned, the cyclic is acceleration-controlled, so there is no need for the pressure to be sustained. Continuing to hold the cyclic forward will just cause the helicopter to continuously get faster, while the helicopter will continue to move forward, even if the cyclic is released. Because the cyclic functions by tipping the rotor disk and redirecting some lift to create thrust, the altitude may drop slightly while the cyclic is applied, but as long as the pressure applied to the cyclic is light, the lost altitude should be negligible and return once the cyclic is put back into the neutral position.

Once the helicopter has begun to move, it will continue to do so for quite some time. During this time, the reference points should be frequently checked to make sure that the two points—the destination and the helicopter—still form a straight line. If not, that means that the helicopter has drifted, and use of the cyclic may necessary to correct the drift. Heading will also need to be maintained; although, barring crosswind conditions, the only time a heading is at risk of changing is whenever the cyclic is applied.

As the target is reached, rearward cyclic pressure should be applied to slow the helicopter down. Releasing and reapplying pressure may be necessary to control the helicopter's airspeed during the approach, with the goal being to slow down to a hover just as the position is reached. The cyclic may be used to increase speed or reverse it, if the destination is under-shot or over-shot respectively, but when making this sort of minor adjustment, the cyclic should be used sparingly in order to maintain control of the helicopter's motion.

Note that even though this is called *forward flight*, the same technique is used when performing *sideward* or *rearward* flight, with only a few modifications. Rather than applying forward cyclic pressure, a sideward or rearward flight requires a sideward or rearward pressure, respectively. Reference points will still be used to watch for signs that the helicopter is drifting from its path, but because of the reduced visibility of the helicopter's side and rear arcs, clearing the area first is especially important, and using ground personnel to assist is highly recommended whenever possible.

Lastly, a problem unique to sideward flight is that the aircraft will attempt to *weathervane*. This is when the helicopter begins to turn due to the helicopter's tail striking the air. In this instance, use of the anti-torque pedals is necessary to counteract the weathervane effect.

Takeoff from Hover

In a normal *takeoff from hover*, the goal is to bring the helicopter from a hovering position to straight and level flight in the air. While similar in principle to forward flight, takeoff differs primarily in the altitude and speed of flight. For safety reasons, certain combinations of velocity and altitude are prohibited, and although the exact range varies depending on the helicopter, a higher velocity generally requires a higher altitude and vice versa. What this means is that while the helicopter is accelerating, it will also need to climb in order to stay within those parameters. Thus, the collective and cyclic must be used in unison. Typically, this maneuver is broken up into five transitional phases, as seen below, starting with a hovering position at phase one and ending with straight and level flight at phase five.

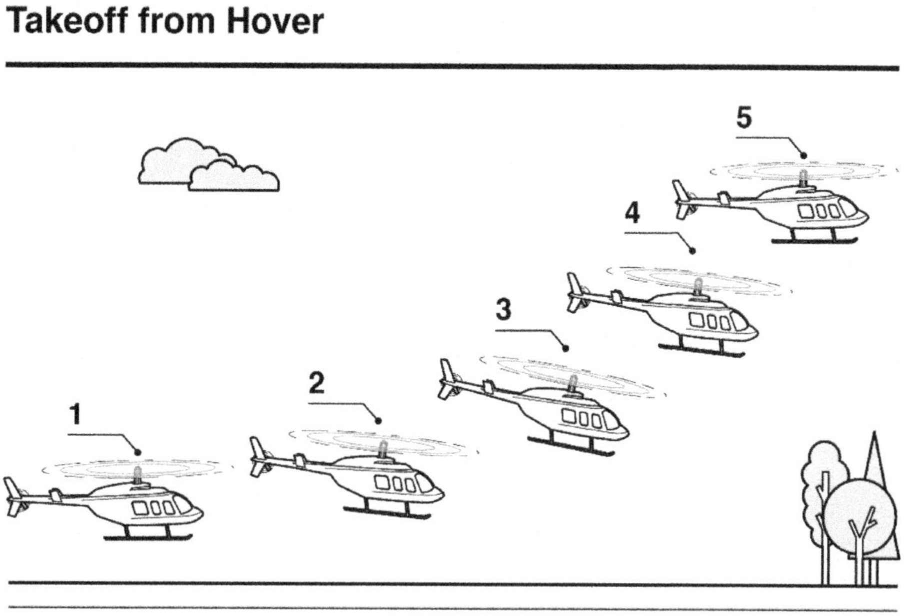

- In *phase one*, the helicopter is in hover. Before moving forward with the rest of the maneuver, a *performance check* is advisable—checking the power required to hover against performance planning data. A performance check is used to determine if the aircraft is within the maximum gross weight and if sufficient power is available to perform the mission.

  The pilot must ensure that all the controls move freely, that they do not get stuck at any point during their full rotation of movement, and that there is sufficient power available to continue. Just as with forward flight, two reference points should be obtained in between the helicopter and its intended direction of flight, which will be used to ensure the helicopter is kept laterally stable.

- The *second phase* begins by applying forward pressure on the cyclic. Because a higher speed is desired, more pressure can be applied to the cyclic than that used when performing forward flight, although the increase should be steady rather than rapid, to ensure the helicopter remains in controlled flight. This greater pressure on the cyclic results in a greater portion of lift being used to create thrust, and while this amount is negligible in forward flight, that is not be the case here.

  The collective will need to be raised to keep the helicopter from losing altitude. Raising the collective requires a proportionate increase in power to keep RPM constant, and the increased power from the throttle will create more torque, which needs to be accounted for with the anti-torque pedals. This means that during this phase, *all four controls need to be used simultaneously*. This is the reason why the cyclic should be increased slowly during this phase, to give ample time to adjust the other controls as needed.

- In *phase three*, the helicopter has gained enough speed to reach *effective translational lift* (ETL). This typically occurs around speeds of 16 to 24 knots; it is easily recognizable by the helicopter beginning to climb and the nose beginning to rise. At this point, the collective should be increased to begin climbing, and the cyclic must be pushed forward enough to counteract the nose's inclination to rise during this phase.

  Again, because the collective is being raised, the pilot should expect to increase the throttle and use the anti-torque pedals to counteract the increased torque and maintain the RPM and heading.

- In *phase four*, the helicopter is climbing and accelerating forward simultaneously. Care must be taken to control both, so that the helicopter does not end up in the shaded area of the height-velocity diagram, but otherwise this phase is mostly about maintenance.

A Sample Height Velocity Diagram for Smooth, Level, Firm Surfaces

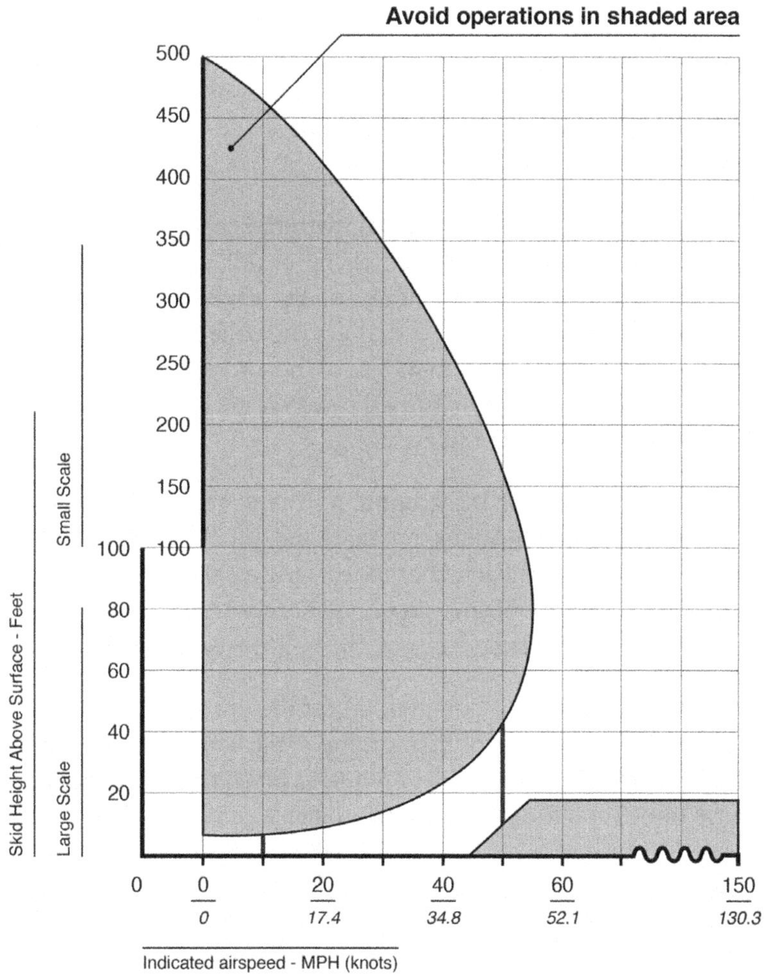

- In *phase five*, the helicopter has escaped the curved section of the height-velocity diagram, and the helicopter can now climb or accelerate, mostly at will. The shaded areas must still be avoided, but typically by this phase, enough distance has been put between the shaded areas and the helicopter's height/velocity that such concerns should only be an afterthought—something to be remembered if airspeed or altitude decreases significantly, but otherwise, it's not likely to be relevant in normal flight. At this point, the helicopter has entered straight and level flight.

Takeoff from Surface

A *takeoff from the surface*, sometimes called a *running takeoff*, is similar in many ways to a takeoff from a hover—both involve increasing altitude and airspeed significantly, and both end up in straight and level flight. In fact, after a certain point, the technique for both become identical. However, there are enough differences, particularly early in the maneuver, to warrant a brief discussion.

Typically, this maneuver is used in preference to a takeoff from a hover only when the helicopter lacks sufficient power to hover. This is usually because the helicopter is overweight, but is also sometimes due to environmental conditions, such as high-density altitude.

As the name implies, a takeoff from the surface involves starting from a landed position and ending in a straight and level flight. This maneuver involves sliding across the surface as speed is picked up, so wheeled helicopters perform better than skidded ones. However, special "skid shoes" can be equipped to the helicopter to help.

The maneuver begins with the helicopter on the surface, with the collective and throttle at their lowest respective settings. Since this maneuver involves the helicopter riding across the ground for some distance, it is especially important to clear the area that will be servicing as the runway. The throttle should be increased until the proper RPM is achieved; then, the collective should be raised. The goal is to reach the point where the helicopter is light on its skids. The anti-torque pedals and cyclic are used as needed to ensure that the helicopter is properly turned towards the intended flight path; then, the pilot should begin with slight forward pressure on the cyclic.

The helicopter will begin to slide forward on the ground. As it does so, use of the anti-torque pedals should continue to keep the helicopter's heading constant, and left or right cyclic should be applied if the helicopter starts to drift laterally. As the helicopter accelerates, the rotor will become more effective, and the helicopter will start to become progressively lighter. This will reduce the friction between the wheels (or skids) and the ground, causing the helicopter to accelerate even faster.

Eventually, the helicopter will become light enough to lift off the ground. As it does, the pilot should increase the collective to assist and prepare to increase the cyclic when the helicopter's nose begins to rise due to translational lift. From this point, the technique to finish takeoff is identical to a takeoff from hover (starting from phase three) and the same procedure can be used.

Straight and Level Flight

In *straight and level flight*, the goal is to keep a helicopter in cruising flight on a straight flightpath and prevent any change in altitude, velocity, or heading that the pilot doesn't initiate. A helicopter is anything but stable, so this is often easier said than done. Fortunately, the helicopter is very controllable, so an experienced pilot can often make corrections before the helicopter has even begun to move, relying more on feel than sight.

The airspeed of a helicopter depends primarily on its attitude (or pitch up or down relative to the horizon), which, in turn, is controlled by the cyclic. Increasing the pressure to the cyclic increases the rate at which the attitude changes, but in most cases, there's little need for rapid acceleration. Instead, the cyclic should be applied slowly and in moderation. Although drag dampens the helicopter's acceleration somewhat, the cyclic is still an acceleration control, and too much forward pressure can make maintaining a constant rate difficult.

Additionally, there is a bit of control lag with the cyclic, in that there is a short delay between input through the cyclic and output through the helicopter. One all too common mistake is for a pilot to adjust the cyclic, believing that nothing has happened, and adjust the cyclic further. This results in the helicopter moving much faster than intended, a loss of altitude, and many other potential complications that could easily be avoid simply by remembering the delay between input and output.

The same procedure also works in reverse, should the speed need to be decreased. Applying aft pressure to the cyclic raises the helicopter's nose and levels out its attitude. With less thrust being generated, the drag created by the helicopter's movements eventually slows the helicopter down until the thrust and the drag are once again opposite. Additional care must be taken while slowing down, however, so that the helicopter does not end in the shaded area of the velocity-height diagram.

Whenever the cyclic is used, the collective almost certainly needs to be adjusted as well. As mentioned previously, the cyclic functions by changing the angle of the rotor disk, which changes the direction of lift and repurposes some of the lift as thrust. In order to keep the altitude the same, lift must remain the same. This can be achieved by increasing or decreasing the collective, while simultaneously adjusting the cyclic, so that the lift lost to thrust is immediately regained by raising the collective. Of course, increasing the collective requires more power, so the throttle also needs to be rotated to keep RPM. More power creates more torque, so the anti-torque pedals need to be used as well.

Banking Turns

There are many parallels between hovering forward flight and straight and level flight. Though each has their own nuances, many of the same principles from one apply to the other as both use a similar mechanism to create and control airspeed. The same cannot be said for hovering turns and banking turns, which share only one commonality—the goal to change the heading of the helicopter.

In a hovering turn, the primary control used to make the turn is the anti-torque pedals. However, when attempting to perform a *banking turn*, the cyclic is used instead. Specifically, the cyclic is pushed laterally to bank the helicopter, which allows the generated thrust to rotate the helicopter in an arc.

The final banking angle, and the rate at which it banks, depends on the duration and force applied to the cyclic. Like always, the control should be applied slowly so that control of the helicopter's turn can be maintained. When the desired turn angle is reached, releasing the cyclic to its neutral position will keep the helicopter banking at that angle.

The greater the angle, the sharper the turn, but this sharpness comes at the cost of the lift generated. Thus, the greater the angle, the more the collective needs to be raised to keep altitude throughout the maneuver and the more power that is needed to keep up the RPM. Care should be taken to ensure that sufficient power is available to complete the turn at the desired angle or else the helicopter will lose altitude.

The anti-torque pedals are still needed during this maneuver, but since the heading is supposed to change during the maneuver, using the helicopter's facing as a guideline for when to apply the pedals will not work. Instead, the pilot should look for signs of slipping or skidding and react appropriately. Both involve a failure of the helicopter's heading and actual direction of movement to match, which result in inefficient flight.

- *Slipping* occurs when the helicopter's banking angle is too great for the rate of turn. In this case, the helicopter's nose will be pointing outwards from the direction of the turn. This results in the helicopter sliding laterally into the turn. This can be corrected by increasing the pressure on the pedal in the direction of the turn or decreasing the pressure on the pedal away from the turn.

- *Skidding* occurs when the helicopter's rate of turn is too great for the angle of bank. In this case, the helicopter's nose will be pointing inwards from the direction of the turn. This results in the helicopter sliding laterally away from the turn. This can be corrected by increasing the pressure on the pedal away from the turn or by decreasing the pressure on the pedal in the direction of the turn.

Once the turn nears the finish, cyclic pressure should be applied in the opposite direction of the turn to level out the helicopter, and the helicopter will stop turning. Reducing the banking angle will restore the lift, so the collective and throttle will need to be decreased. Otherwise, the helicopter might climb. Also, if the cyclic is not applied until the turn is actually reached, the helicopter will end up turning too far as the helicopter requires time to level out. To prevent this, the pilot should begin to apply counter cyclic pressure shortly before arriving at the desired angle.

Normal Climb and Descent

Although technically two separate maneuvers, climbing and descending are very similar in execution, with one being basically the opposite of the other. Both primarily use the collective and the throttle, although both allow for cyclic adjustments to influence the rate of climb and descent at the expense of airspeed.

To perform a climb, the collective must be raised to generate additional thrust. The throttle needs to be raised in tandem to keep RPM consistent, which requires additional application of the anti-torque pedals to keep the helicopter's heading constant. Since this application increases the total thrust generated (irrespective of the angle of the rotor disk), it will also increase the airspeed of the helicopter, if the helicopter has any forward attitude. Minor aft pressure should be applied to the cyclic to raise the helicopter's attitude slightly. This change in attitude will decrease the amount of thrust derived from the lift, but the lost thrust should be replaced as the net lift is increased.

If a faster climb is desired, more aft pressure can be applied to the cyclic, further reducing the amount of generated thrust to increase the amount of generated lift. This results in decreased airspeed, however, so caution must be taken to ensure that the helicopter continues to maintain enough airspeed to stay out of the shaded areas of the velocity-height diagram. However, as altitude increases, the shaded area gets smaller and smaller, allowing progressively lower speeds.

When leveling off a climb, the helicopter's momentum will carry it beyond the desired altitude, so the pilot should begin leveling off a bit before reaching the desired height. Typically, 10% of the helicopter's rate of ascent is ideal, so a 100 FPM climb would only need 10 feet to stop in time, while climbing at 700 FPM would need 70 feet.

The collective should be lowered slowly, adjusting the throttle and anti-torque pedals as needed to keep the RPM and heading consistent throughout the process. As the lift decreases, the thrust decreases. The pilot should apply forward pressure on the cyclic as well to keep airspeed and assist with reducing lift. As the intended height is reached, cyclic pressure should be applied to return to the desired attitude, and any final adjustments to the collective and throttle should be made to stabilize the helicopter's altitude.

A descent functions very much the same, just in reverse. To descend, thrust must be decreased, to allow gravity to pull the helicopter down. To do this, the collective must be lowered and the throttle dialed back. With less lift being generated, thrust and airspeed will also decrease. If a constant airspeed is desired, the cyclic will need to be pushed forward slightly to increase the thrust component of the lift. Since the greater thrust component comes at the cost of the lift, this also results in the helicopter descending faster.

By applying the same method, the descent can be sped up by increasing the forward pressure on the cyclic further and increasing airspeed, causing the helicopter to descend more rapidly. Just as with a climb, the helicopter must not be allowed to reach the shaded area of the velocity-height diagram.

When leveling off a descent, just as with a climb, a pilot should expect forward momentum to carry the helicopter some distance after beginning to slow down. The same 10% rule applies here. To begin leveling off from a descent, the collective should be raised slowly, controlling throttle and the anti-torque pedals to maintain RPM and heading, just as was done when climbing.

Also like climbing, leveling off will influence airspeed; in this case, the increasing lift will increase thrust, and aft pressure to the cyclic is needed to keep airspeed constant. Adjustments should continue to be made to the collective and cyclic, until the helicopter has leveled out at the desired airspeed.

Taxiing

Taxiing is less a maneuver and more of a combination of other maneuvers performed with additional limitations. Taxiing is primarily used to move a helicopter a short distance, either to ferry passengers or cargo or simply to reposition the helicopter for some other purpose, such as a cleared runway to allow the helicopter to take off. Taxiing is also unique in that it often involves the use of taxiways or other prescribed routes to follow ahead of time, not unlike a road. There are three types of taxiing, each with its own uses. The three are hover, air, and surface taxiing.

- *Hover Taxiing* involves moving about in a hovering helicopter and employs hovering, forward flight, and hovering turns to move about. Although the exact height of flight varies by taxiway, it is virtually always no more than 25 feet above the ground. Because of the low speed of hover taxiing, this is most often used to reposition a helicopter rather than as a form of transportation.

- *Air taxiing* involves higher speeds and altitudes, although generally it does not exceed heights of 100 feet. Since the helicopter is flying both higher and faster, straight and level flight and banking turns are used instead, as well as climbing and descending when necessary.

  Additionally, a takeoff is always required to enter an air taxi, whether from a hover or the surface. Otherwise, there is little difference between free flight and a taxiway, though in the latter, the helicopter pilot is expected to avoid flying directly above other vehicles or personnel, whether on the ground or in the sky.

- *Surface taxiing* is the only one that requires any technique that hasn't already been discussed. In a surface taxi, the helicopter is kept on the ground the entire time. To accomplish this, the collective is raised until the helicopter is light on its wheels (or skids), just as if preparing for a surface takeoff. Also like a surface takeoff, the collective is then used to create forward movement. That, however, is where the similarities end.

  When performing a surface taxi, the helicopter should never move faster than a brisk walk, to prevent the lift from having any chance of taking the helicopter off the ground. Additionally, although some cyclic is needed to start the forward movement, the speed of the helicopter is controlled through the collective instead.

Although the friction between the wheels (or skids) and the ground makes the helicopter far more stable than when it is in the sky, yawing still occurs, requiring some anti-torque pedal application to negate. Likewise, any strong winds capable of pushing the helicopter need to be countered by applying cyclic pressure in the direction of the incoming wind.

Practice Questions

1. When generating lift, what should the pressure below the airfoil be in relation to the pressure above the airfoil?

   a. Greater than it
   b. Less than it
   c. Equal to it
   d. The opposite of it
   e. The pressures above and below do not matter.

2. What is the load factor of a 2000-pound helicopter that is currently experiencing an effective load of 4000 pounds?

   a. 0.25
   b. 0.5
   c. 1
   d. 2
   e. 4

3. Which of the following type or types of drag increase as the speed increases?

   I. Parasitic drag
   II. Induced drag
   III. Profile drag

   a. I only
   b. II and III
   c. I and II
   d. I and III
   e. I, II, and III

4. Which of the following decreases upon an increase to the collective pitch controller?

   a. Angle of incidence
   b. Altitude
   c. Drag
   d. Lift
   e. RPM

5. Which of the following is FALSE about the governor?

   a. It adjusts the throttle as needed.
   b. It uses an electronic sensor.
   c. It controls the pitch of the rotors.
   d. It regulates the power to maintain RPM.
   e. It requires almost no manual control.

6. If the cyclic pitch control is pushed to the left, which way will the rotor disk tip and which way will the helicopter move?
   a. Right; right
   b. Right; left
   c. Left; right
   d. Left; left
   e. Backwards; forwards

7. Which direction(s) are the anti-torque pedals are designed to move the nose?
   a. Up and down
   b. Left and right
   c. Forward and backwards
   d. The anti-torque pedals move the nose in all of the above directions.
   e. The anti-torque pedals do not move the nose.

8. Which of the following is NOT included in a helicopter's base empty weight?
   a. The pilot
   b. The engine
   c. Optional equipment
   d. Transmission oil
   e. The rotors

9. Which of the following are dangers of overloading a helicopter?
      a. Decreased performance
      b. Structural damage
      c. Shifted center of gravity
   a. II only
   b. I, II, and III
   c. I only
   d. I and III
   e. I and II

10. If the helicopter is nose down while attempting to hover, and there are no winds, then it most likely has which of the following?
   a. Front-loaded center of gravity
   b. Aft-loaded center of gravity
   c. Balanced center of gravity
   d. A changing center of gravity
   e. The center of gravity cannot be determined with the information provided

11. When performing a vertical takeoff to a hover, what should be done if the heading begins to change?
   a. Quickly adjust the cyclic pitch control
   b. Slowly adjust the cyclic pitch control
   c. Quickly adjusting the cyclic pitch control and the anti-torque pedals.
   d. Quickly adjust the anti-torque pedals
   e. Slowly adjust the anti-torque pedals

12. Which of the following would make a good reference point when attempting to hover?
   a. The tail rotor
   b. Another helicopter
   c. A building
   d. A moving car
   e. The main rotor

13. Which of the following should not need to be maintained while performing a hovering turn?
   a. Altitude
   b. RPM
   c. Heading
   d. Position
   e. Center of gravity

14. Which of the following is a complication unique to sideward hovering flight, as opposed to forward and rearward flight?
   a. Increased clearance needed due to dipping of the tail during the maneuver
   b. Additional pressure needed on the anti-torque pedals to deal with the weathervane
   c. Drastically-reduced visibility
   d. Decreased RPM due to the increased pitch of the rotor blades
   e. None of the above

15. When performing an air taxi, which of the following should be avoided?
      I. Flying in the shaded area of the height-velocity diagram
      II. Flying over other aircraft, vehicles, and personnel
      III. Flying in crosswind conditions
   a. I only
   b. I and II
   c. I and III
   d. II and III
   e. I, II, and III

16. Which of the following does not need to be checked *before* beginning a normal takeoff from hover?
   a. Power
   b. Balance
   c. RPM
   d. Flight controls
   e. Fuel levels

17. In straight-and-level flight, which control is primarily used to increase airspeed?
   a. The wind speed
   b. The cyclic pitch control
   c. Anti-torque pedals
   d. The throttle
   e. The collective pitch control

18. A helicopter pilot is attempting to take off when she notices that the helicopter's nose tilts down after leaving the ground. What is most likely the problem?
    a. Too much main rotor torque
    b. Imbalanced center of gravity
    c. Insufficient power
    d. Gremlins
    e. Changing winds

19. A helicopter begins a descent at 300 feet above the ground, with the intent to lower to 200 feet above the ground. The pilot begins the descent at a steady rate of 400 FPM. At what altitude should he begin leveling out the helicopter?
    a. 200 feet
    b. 220 feet
    c. 240 feet
    d. 260 feet
    e. 300 feet

20. Which of the following would NOT be included in the maximum internal weight?
    a. The pilot
    b. Cargo stored in the aft of the helicopter
    c. Useable fuel
    d. Passengers
    e. An object supported by a winch

21. A pilot wishes to reposition the helicopter while hovering to get to a designated runway for a surface takeoff. After performing a hovering turn to set her bearing properly, the pilot needs to move the helicopter forward. To do this, she should do which of the following?
    a. Increase the collective
    b. Briefly apply pressure to the cyclic, then release the cyclic
    c. Apply gentle and constant pressure to the cyclic
    d. Progressively increase the throttle
    e. Briefly increase the throttle, then release the throttle

22. While performing a banking left turn, the pilot notices that the helicopter has begun to skid. What can he do to get out of this?
    a. Decrease pressure on the left anti-torque pedal or increase pressure on the right anti-torque pedal
    b. Decrease pressure on the right anti-torque pedal or increase pressure on the left anti-torque pedal
    c. Apply additional leftwards cyclic pressure
    d. Apply additional rightwards cyclic pressure
    e. Increase the throttle

23. A pilot has successfully identified that the helicopter has an aft-loaded center of balance. Which of the following methods would correct this?
   a. Adding cargo to the back of the helicopter
   b. Apply additional forward cyclic
   c. Increase the throttle
   d. Add additional fuel
   e. Removing cargo from the back of the helicopter

24. A helicopter performing a takeoff from hover has just entered effective translational lift and has begun to increase the collective. Which step of the takeoff is the helicopter currently in?
   a. Step one
   b. Step two
   c. Step three
   d. Step four
   e. Step five

25. A pilot is performing a straight and level flight when he realizes that the helicopter has drifted to the left of the intended path. What should he do?
   a. Apply pressure to the right anti-torque pedal
   b. Apply rightward pressure to the cyclic
   c. Remove cargo from the back of the helicopter
   d. Speed up the helicopter to reach the destination before the helicopter drifts further
   e. Nothing, drift is not a problem as long as the area is clear

26. Which of the following would increase the torque on the helicopter?
      I. Applying pressure on the cyclic
      II. Decreasing the collective
      III. Increasing the throttle
   a. I
   b. II
   c. III
   d. II and III
   e. All of the above

27. Which of the following is true about induced drag?
   a. It is strongest when the helicopter is moving at higher speeds.
   b. It is created by the airfoil striking the air.
   c. It always pushes in a direction opposite of lift.
   d. All of the above are true.
   e. None of the above are true.

28. If sufficient takeoff power is in doubt, what should the pilot do?

I. Delay the takeoff until the aircraft is lighter
II. Attempt the takeoff anyway to check power
III. Wait until the density altitude has decreased

a. I only
b. II and III
c. I and II
d. I and III
e. All of the above

29. Which of the following is not one of the four fundamental maneuvers?

a. Turning
b. Straight and level flight
c. Climbing
d. Takeoff
e. Descending

30. A pilot notices that as the helicopter flies, the nose slowly drops. The cyclic is in its neutral position. What is most likely the cause of this?

a. The fuel in the rear of the helicopter is being used up, slowly shifting the center of balance forward.
b. The cyclic is improperly calibrated and tipping the helicopter even into the neutral position.
c. The helicopter is entering a higher density air front, pushing the nose down.
d. The helicopter's tail rotor is creating too much torque, pushing the tail up and the nose down.
e. It cannot be determined without additional information.

31. Lowering the collective while the helicopter is flying straight and level will have what effect, assuming everything else remains constant?

a. It will cause the nose to pitch down, the helicopter's airspeed to increase, and will produce left yaw.
b. It will cause the nose to pitch up, the helicopter's airspeed to decrease, and will produce right yaw.
c. It will cause the nose to pitch down and the helicopter's airspeed will decrease.
d. It will have no appreciable effect on the airspeed of the helicopter.
e. There is not enough information provided to make a determination.

32. While hovering, a pilot wishes to increase her altitude, so she uses the collective. Which control will the pilot NOT need to use while doing so, assuming ideal conditions?

a. The cyclic
b. The anti-torque pedals
c. The throttle
d. All the above are needed while raising the collective
e. None of the above are necessary while raising the collective

33. After achieving the desired altitude, the pilot decides to perform a takeoff from a hover and applies forward pressure on the cyclic to begin forward movement. Which control/s will he need to use while doing so, assuming ideal conditions?

I. The collective
II. The anti-torque pedals
III. The throttle

a. II only
b. I and II
c. I and III
d. II and III
e. I, II, and III

34. Although both parasitic drag and profile drag increase as speed increases, which of the following correctly identifies the difference between the two?

a. Profile drag starts out as more significant than parasitic drag and remains more significant at all speeds.
b. Parasitic drag starts out as more significant than profile drag and remains more significant at all speeds.
c. Profile drag starts out as more significant, but parasitic drag becomes more significant at higher speeds.
d. Parasitic drag starts out as more significant, but profile drag becomes more significant at higher speeds.
e. Parasitic drag and profile drag are both equally significant

35. When performing a hovering turn on a windy day, at what point is weathervane most likely to cause problems?

a. When the helicopter's heading is into the wind
b. When the helicopter's heading is with the wind
c. At the end of the turn
d. At the beginning of the turn
e. The risk of complications is equal throughout the turn

36. When climbing at cruising speed, in addition to raising the collective, the rate of climb can be increased by which of the following actions?

I. Applying rearward cyclic pressure to reduce airspeed
II. Increasing the throttle to increase power
III. Applying forward cyclic pressure to increase airspeed

a. I only
b. II only
c. III only
d. I and II
e. I and III

37. Which of the following maneuvers should be avoided when flying with a heavy load?

I. Large angle banking turns
II. High speed straight and level flight
III. Takeoff from hover

a. I
b. II only
c. III only
d. II and III
e. I, II, and III

38. Which of the following influences the G load?

I. The weight of the helicopter
II. The weather
III. Air density

a. I only
b. I and II
c. I and III
d. II and III
e. I, II, and III

39. At what pace should a helicopter in forward flight move?

a. About as fast as a brisk walk
b. No faster than jogging speed
c. As fast as desired as long as control can be maintained
d. Not faster than a run, and even slower if the area is uneven
e. About as fast as a car on the highway

40. Which control is a rate-controlled instrument?

a. The collective
b. The throttle
c. The cyclic
d. The anti-torque pedals
e. All of the above are rate-controlled instruments

Answer Explanations

1. A: The pressure below the airfoil needs to be greater to push upwards on the airfoil. If the pressures are equal, there will be no lift generated at all, and if the pressure above is greater, the pressure from above would actually push down on the airfoil, generating negative lift.

2. D: Since the effective load is double its normal weight, the load factor is 2. In this case, simple eyeballing of the numbers is likely enough to solve this problem. If the numbers were less convenient, the problem could still be solved by using the equation, $G=L_1/L_0$. Using 4000 as L_1 and 2000 as L_0, the result would be G=4000/2000, which equals 2.

3. D: Parasitic drag and profile drag both increase as airspeed increases. Induced drag decreases as airspeed increases. Both parasitic drag and profile drag are created by air resistance, which increases the faster the helicopter is going. However, induced drag is created by backwash from the main rotor and thus, it depends only on the angle of attack of the helicopter, which is typically lower in higher speed flight.

4. E: Raising the collective pitch control increases the pitch of the blades, which increases the angle of incidence. This generates lift, which results in increased altitude and drag, which then reduces the RPM, if the throttle is not simultaneously increased to compensate.

5. C: The governor utilizes an electronic sensor, which is fully-automated to maintain the RPM of the rotor blades. The governor does not, however, control the pitch of the rotors. That function is reserved for the collective pitch control and cyclic pitch control.

6. D: The tilt of the disk and the direction of horizontal movement are both directional with the cyclic pitch control. Therefore, both will go in the same direction that the cyclic pitch control is pushed, which, in this case, is left.

7. B: The anti-torque pedals allow the helicopter to alter its heading left and right. The pedals cannot move the nose up and down, nor can they create forward or backwards thrust. The cyclic is the instrument that controls the helicopter's attitude and is responsible for creating the thrust that moves the helicopter forward and backwards.

8. A: All of the listed items are included in a basic empty weight except for the pilot. Had the question asked about the licensed empty weight, both *A* and *D* would have been correct, as the transmission oil is not included in that case.

9. E: While overloading a helicopter can cause a shifted center of gravity (if the helicopter is not properly managed), that is not a direct effect of overloading a helicopter. Structural damage and decreased performance, however, are both potential effects of overloading, regardless of how the weight is distributed.

10. A: A nose-down position means that there is more weight located near the front of the helicopter; thus, the helicopter is front-loaded. Had the helicopter been aft-loaded, the nose would have pointed up, and a centrally-balanced helicopter would be parallel with the ground.

11. E: If the heading is moving, the anti-torque pedals are used to compensate. The cyclic pitch control would be used to counteract a horizontal shift in position, rather than heading. Additionally, like all controls, the anti-torque pedals should be adjusted slowly to avoid over-controlling.

12. C: A moving object like a car would make a poor point of reference as it would be difficult—if not impossible—to figure out whether its shift relative to the current position is a result of its movement on the ground or the movement in the air. Likewise, using another helicopter as reference, even if that helicopter is also in hover, is ill-advised, as there is no guarantee its position is stable. Using one's own main rotor or tail rotor would be silly, given that these pieces are attached solidly to the helicopter and will always appear stationary relative to the rest of the helicopter. A building, however, is fixed to the ground, which makes it an excellent point of reference.

13. C: Altitude, RPM, and position should ideally remain constant while performing a turn. Changing the heading, however, is the purpose of the maneuver. Maintaining center of gravity will also help keep the aircraft balanced.

14. B: Increased clearance is only necessary for rearward flight, and though there is some loss of visibility in making sideward motion versus forward motion, this is a more significant issue in rearward hovering flight. While RPM does decrease when adjusting the pitch of the blades, this is true of all maneuvers. Only the weathervane effect of the tail is truly unique to sideward hovering flight.

15. E: All three items should be avoided whenever possible as they can all increase the risk of damage, either to the helicopter itself or to objects around it.

16. C: The power, balance, and flight controls must be checked before takeoff to ensure the helicopter can safely perform the maneuver. Fuel levels should also be sufficient for the planned mission. In contrast, the RPM should not be significantly changing during a hover, although it will need to be checked while performing the maneuver.

17. B: The cyclic pitch control is used to control airspeed. The anti-torque pedals are used to control heading, the throttle controls RPM, and the collective pitch controls altitude. Wind cannot be relied on as a means by which to increase speed.

18. B: A nose-down helicopter is most often the result of a front-loaded center of gravity. Excessive torque results in the helicopter spinning, while insufficient power prevents the helicopter from successfully taking off.

19. C: By using the 10% rule, it is clear that leveling out at 240 feet is ideal. If the helicopter does not begin leveling out until 220 or 200 feet, the helicopter will almost certainly overshoot the target, while stopping too early may result in ending short of the desired destination.

20. E: The pilot, the cargo, passengers, and the fuel are all inside the helicopter and are counted towards the maximum internal weight. An object supported by a winch, however, is outside the helicopter and is instead applied to the maximum external weight.

21. B: The cyclic is an acceleration control, and holding the cyclic down will cause the helicopter to continuously accelerate. Since the helicopter is in a hover, and high speeds should be avoided while performing forward flight, the correct method is to apply brief pressure, allowing the helicopter to accelerate to some speed and then release the cyclic to maintain that speed. Increasing the collective would only work to increase speed if the helicopter already had a forward attitude.

22. A: Skidding is caused when the rate of turn is too great for the angle of bank. Therefore, the solution to skidding is to decrease the rate of turn, which requires either less pressure on the pedal in the direction of the turn (in this case, left) or more pressure on the pedal in the opposite direction. While increasing the banking angle (by applying additional left pressure on the cyclic) would also stop the skidding, it would also make the banking angle greater, which would have numerous other effects. For this reason, correcting the anti-torque pedals is the preferred method.

23. E: Removing the offending weight is the simplest way to correct a bad center of balance. While applying forward cyclic would treat the symptoms of the problem, doing so would not do anything to resolve the root of the issue and would still result in reduced performance. The throttle has nothing to do with the center of balance, and given that most fuel reserves are in the rear of the helicopter, adding fuel would likely exacerbate the problem.

24. C: Once effective translational lift has been reached, step two has officially ended, but since the pilot has only just entered ETL and applied the cyclic, the helicopter has not yet reached step four. Therefore, the helicopter is currently at step three.

25. B: Using the anti-torque pedals would help if the nose was drifting to the side, but if the helicopter itself is displaced from the flight path, the anti-torque pedals wouldn't help. Likewise, ignoring the problem or trying to hurry past the problem won't solve anything. Removing cargo will also not help in this situation. Using the cyclic to nudge the helicopter back solves the problem without creating any new ones and is the ideal answer.

26. C: Increasing the throttle will increase the power, which creates more torque. Decreasing the collective would actually decrease the torque, assuming that the throttle was appropriately controlled to keep RPM constant. The cyclic primarily rotates the rotor disk to convert lift to thrust and does not directly increase the torque of the helicopter.

27. C: Parasitic drag, not induced drag, is strongest at higher speeds, and profile drag is created by the airfoil striking the air. Induced drag actually generally gets less potent as airspeed increases. This is created by the vacuum left behind by the airfoils and the air rushing to fill that hole. Induced drag does, however, create the drag in the direction of lift, which is why induced drag's significance depends on the attitude of the helicopter.

28. D: If takeoff power is in question, the pilot should wait until the temperature cools off, and the density altitude will decrease. Also, departures should be planned either in the morning or later in the day. If possible, the load on the aircraft should be reduced or fuel burned off to reduce weight. If the takeoff is attempted and the aircraft does not have sufficient power, it could put the aircraft in an unsafe flight profile.

29. D: Takeoff, while a common maneuver, is not considered a fundamental maneuver, primarily because takeoff is at most performed once per flight, while turning, climbing, descending, and flying straight will be performed constantly.

30. A: While an improperly-calibrated cyclic is possible, something like that would most likely be noticed immediately, and the cyclic somehow getting more uncalibrated during flight would not explain why the pilot saw the helicopter slowly dropping. A higher air density would indeed have an impact on a helicopter's flight, but air density would change at such a slow gradient that the change would be imperceptible to the human eye. If the tail was producing too much torque, the tail would move horizontally, not vertically. Fuel, however, does have weight, and as the fuel is consumed, the weight of the remaining fuel would decrease.

31. A: The nose of the helicopter always follows the direction the collective is moved. For example, if the collective is increased, the nose of the helicopter pitches up. It does the opposite when lowered. Also, when the collective is raised, it takes the left pedal to counteract the torque effect and the right pedal when the collective is reduced.

32. A: When raising the collective, more power will be needed to keep RPM up, which will require raising the throttle in sync. More power to the rotor will result in more torque, which will require using the anti-torque pedals to compensate. The cyclic, however, should not be needed during this maneuver, except perhaps to negate any drift caused by wind. Were the helicopter in forward flight, the cyclic would need to have slight rearwards pressure applied to account for the additional lift creating additional thrust.

33. E: With the cyclic pushed forward, additional lift is created by the rotor disk tipping forward, but this comes at the expense of lift. Thus, the collective will need to be raised to keep altitude up, which will require additional power from the throttle and additional counter torque from the anti-torque pedals.

34. C: Profile drag is fairly consistent at most speeds, and although profile drag does increase somewhat with airspeed, parasitic drag catches up to and then surpasses profile drag in magnitude.

35. B: Weathervaning occurs when the wind blows at the helicopter and strikes the tail, which usually sticks out significantly from the rest of the body. This effect makes attempting to turn the helicopter into the wind (and thus, turning the tail with it) require less power, while attempting to turn the helicopter with the wind (and the tail into it) requires more power. While this makes turning at all points trickier, the effect becomes particularly challenging when the helicopter reaches a heading directly with the wind. At this point, the wind will go from resisting the turn to assisting with the turn, which can result in the helicopter turning much faster than expected. The opposite happens when the heading is into the wind, at which point the wind goes from assisting the turn to opposing it. This will often stall the turn, giving the pilot plenty of time to adjust the controls to accommodate the problem.

36. A: While the throttle will need to be increased as the collective increases to maintain RPM, the throttle should never be used to increase the rotor's RPM beyond the normal range. Applying additional forward cyclic will actually decrease the rate of climb as the greater angle of attack results in greater thrust, but reduced lift. Therefore, decreasing the angle of attack by applying rearward pressure will result in less thrust being generated, but more lift, thus increasing the rate of climb.

37. A: Straight-and-level flight is actually the best place for a heavily encumbered helicopter as the angle of attack is usually lowest there, allowing for less thrust and more lift. Similarly, there is no reason to avoid a takeoff from hover, if the helicopter has sufficient power to do so. The main reason most heavily-laden helicopters use a takeoff from surface instead is that they are too heavy to hover properly; thus, they are unable to perform the maneuver. However, performing a steep banking turn can be very dangerous in a heavy helicopter as the load factor can get rather high, and potentially damage the helicopter or reach a point where the helicopter cannot produce enough power to maintain altitude, resulting in an undesired descent.

38. D: Both weather and air density can affect the efficiency of the air rotor and, thus, the G load. However, while the helicopter's weight does have an effect on the actual effective load on the helicopter, the G load is a multiplicative factor applied to—rather than derived from—the weight of the helicopter to determine the helicopter's effective weight.

39. A: Even in ideal conditions, unless the helicopter is preparing to takeoff or is already in straight and level flight, there is no reason for the helicopter to move faster than a brisk walk.

40. D: The cyclic is an acceleration-controlled instrument, while both the collective and throttle are position-controlled. Only the anti-torque pedals are rate-controlled.

Spatial Apperception

There are 25 questions in the 10-minute Spatial Apperception section of the SIFT exam. Like the Simple Drawings and Hidden Figures sections, the questions in the Spatial Apperception section also assess the test taker's ability to process and interpret visual information. The term *apperception* refers to the process of understanding new information or observations by relying on past experience and knowledge.

The Spatial Apperception questions have been specifically designed to assess one's ability to translate visual information provided in a given perspective into an accurate mental image of the same information from a very different perspective. This is a critical skill for Army pilots and achieving a high score in this section will demonstrate one's readiness to handle the rigor and challenges as such.

Each question depicts a sketch of a pilot's visual perspective of the terrain and horizon from the cockpit of an aircraft. This drawing is followed by five additional drawings (answer choices) that show the aircraft as observed from the perspective of someone on the ground. In each of these choices, the aircraft is depicted in a different attitude. Test takers must determine which of the five drawings correctly depicts the attitude of the aircraft that would provide the pilot in the cockpit with the terrain view illustrated in the first drawing.

When answering the Spatial Apperception questions, test takers are looking to answer the following three questions for each drawing from the pilot's perspective:

1. Is the aircraft flying straight ahead, climbing, or diving? In other words, what is its vertical orientation?

2. Are the aircraft's wings level, or is the aircraft banking right or left?

3. In which direction is the plane heading?

To answer these questions, it is often best to start by determining the aircraft's orientation, which can be ascertained by examining the horizon in the drawing. The horizon is where the sky meets the terrain (either the shaded area depicting the land or the sea). If the horizon is in the middle of the drawing (such that there is an equal amount of space above the horizon as sky and below the horizon as terrain), then the aircraft is level, or flying straight ahead. On the other hand, if most of the drawing is taken up by sky above the horizon, then the aircraft is climbing, while drawings showing very little sky and mostly the terrain below are depicted from the cockpit of an aircraft that is diving.

To determine if the aircraft is flying level or if it banking to either side, the horizon line comes into play. If the horizon line is level, then the wings of the aircraft are level. If instead, the horizon line slopes from the upper left down to the lower right in the drawing, the plane is banking left. The plane is banking right if the horizon line slopes up from the lower left to the upper right. Essentially, the plane is banking in the direction in which the horizon is elevated on that side of the drawing.

The final piece of information needed to determine the correct image is to identify the direction in which the aircraft is headed: toward the land, sea, or along the coastline. The aircraft is flying alongside the coastline if it's oriented at a right-angle relative to the coast. However, if this angle differs from 90 degrees, the aircraft is heading toward the land or the sea (which is determined by the specific image and angle of approach).

It is important to carefully practice these questions and employ significant attention to detail during the Spatial Apperception section on the official SIFT attempt. This can be a deceptively simple section of the exam, so test takers are encouraged to carefully evaluate each drawing using the three-question method and work through all of the practice questions. Candidates often confuse the horizon and coastline, particularly in angled drawings, so test takers should carefully identify what they are looking at in the cockpit perspective image before evaluating the subsequent perspectives. It should be noted that all answer choices for a given question will be from the same sided view of the aircraft—either all from coastline left or all from coastline right. Therefore, it is helpful to scan the answer choices first to determine which of the two views is employed in the perspective of the viewer on shore before evaluating the three questions from the cockpit perspective.

Practice Questions

1.

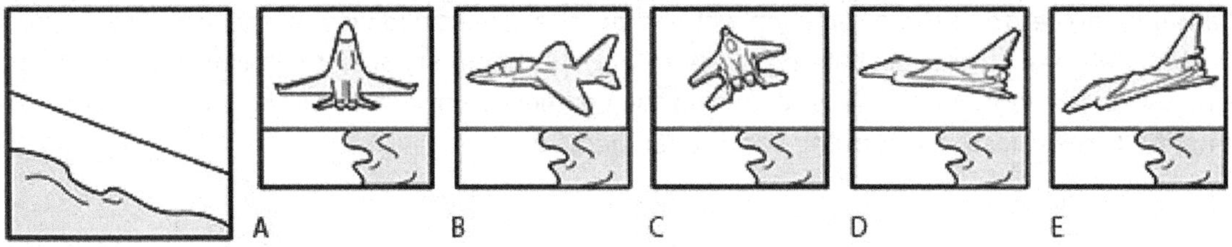

A B C D E

2.

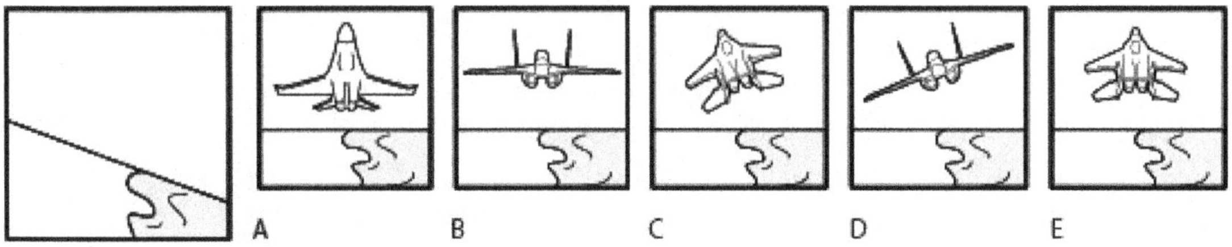

A B C D E

3.

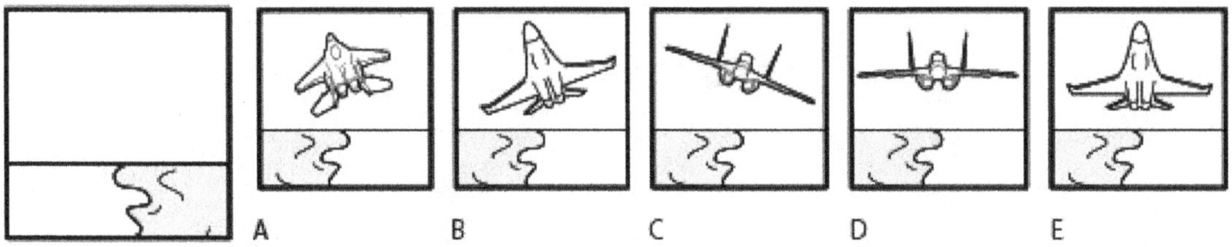

A B C D E

4.

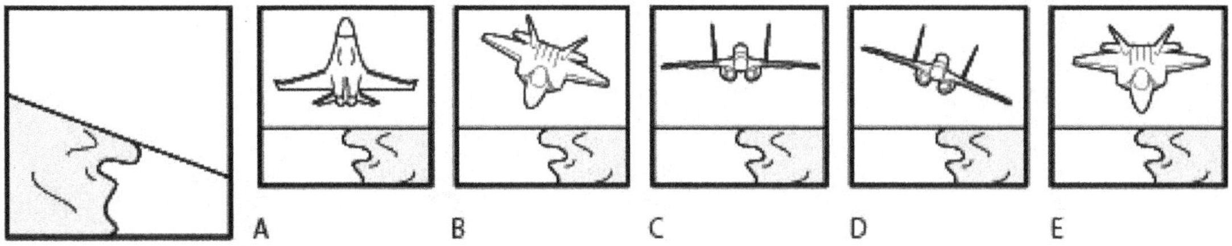

A B C D E

5.

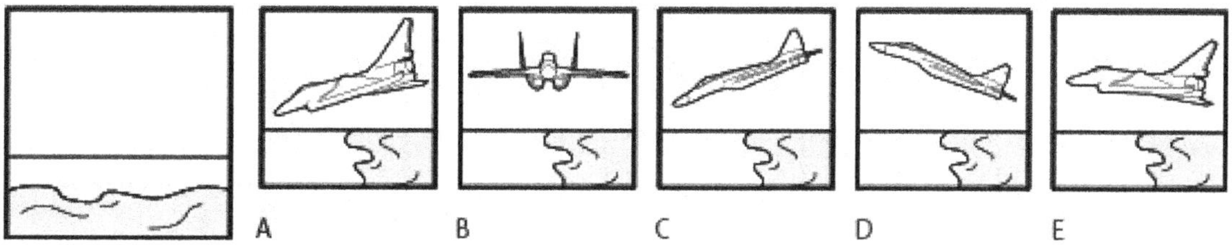

6.

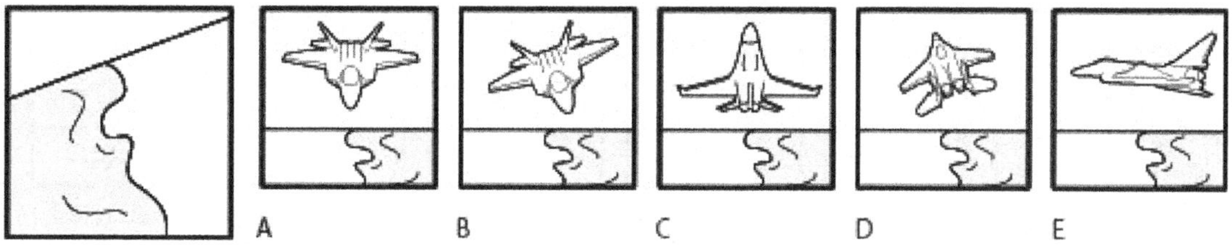

7.

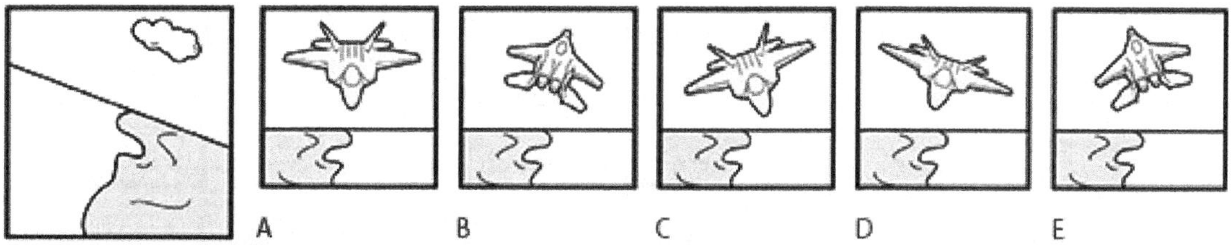

8.

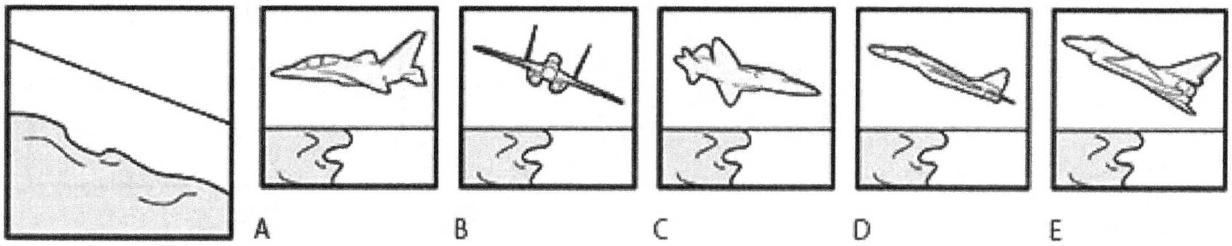

9.

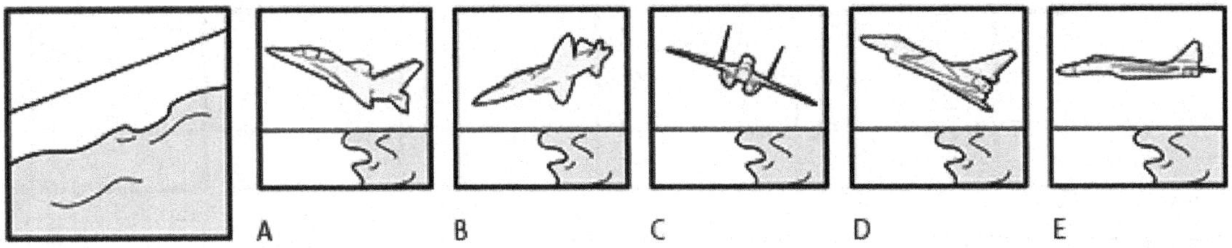

10.

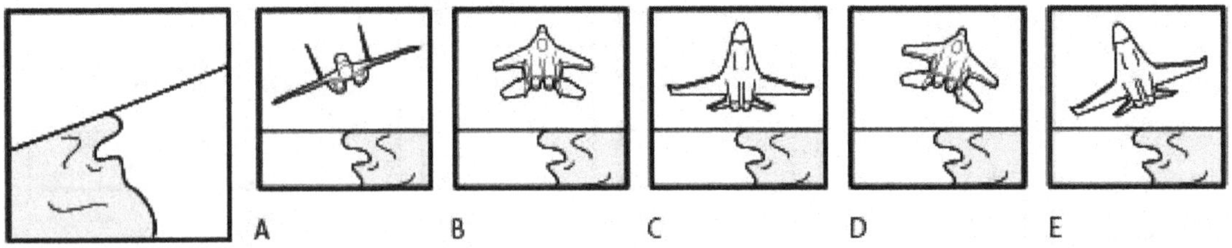

11.

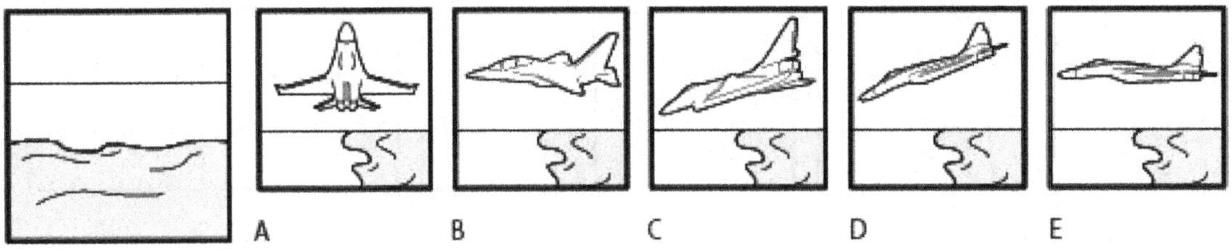

12.

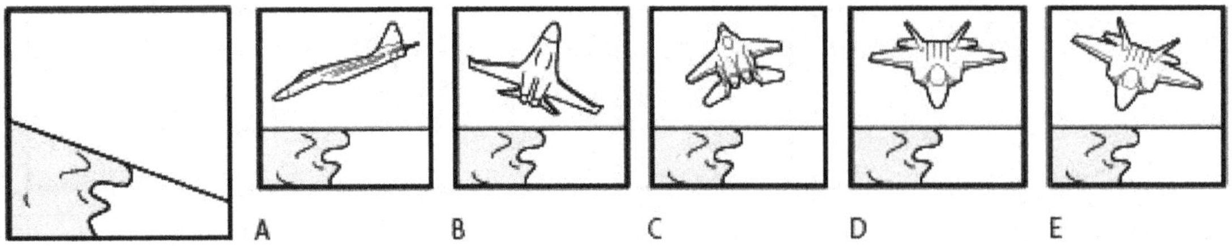

13.

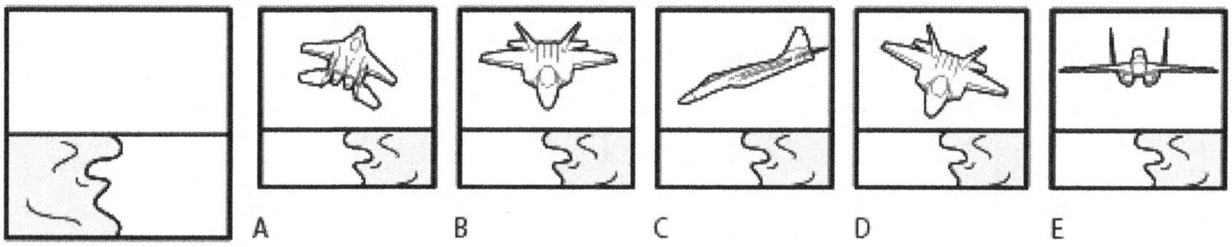

A B C D E

14.

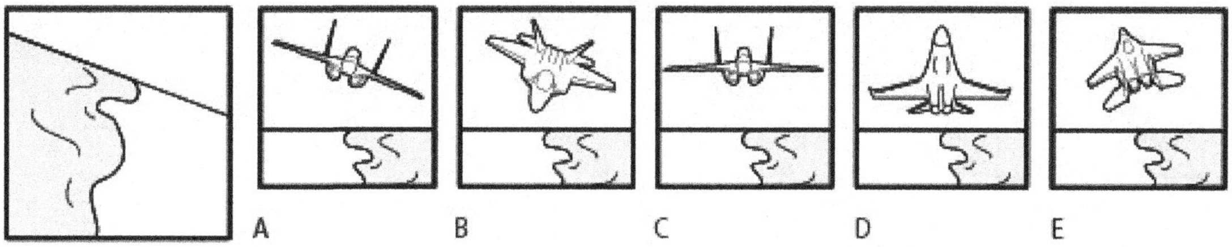

A B C D E

15.

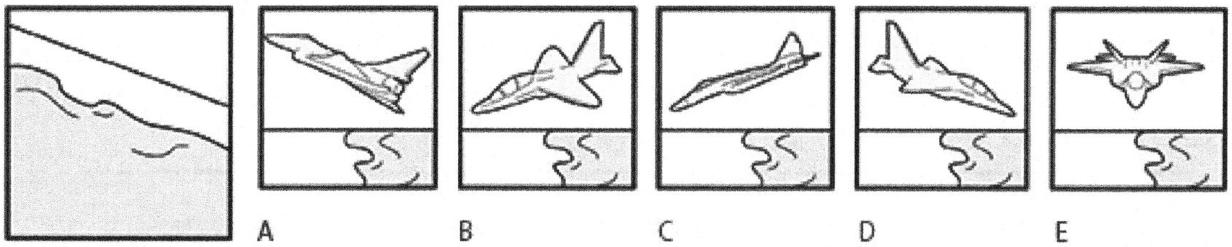

A B C D E

16.

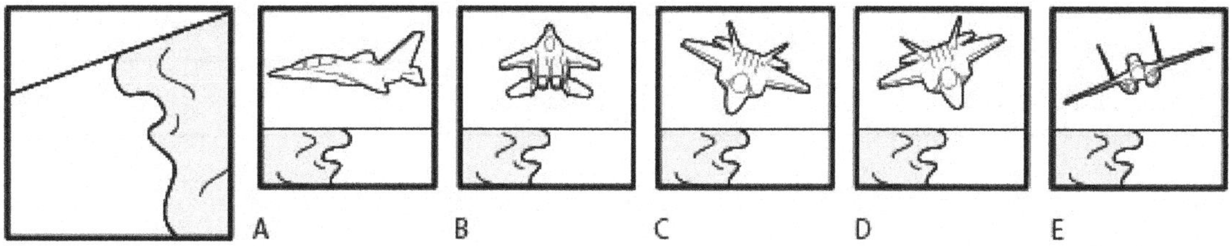

A B C D E

17.

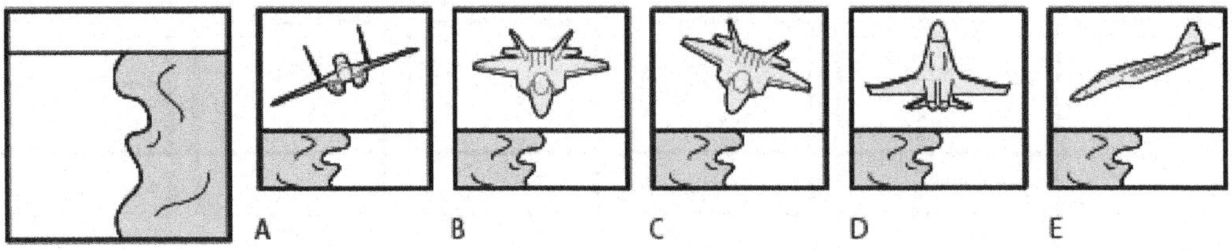

18.

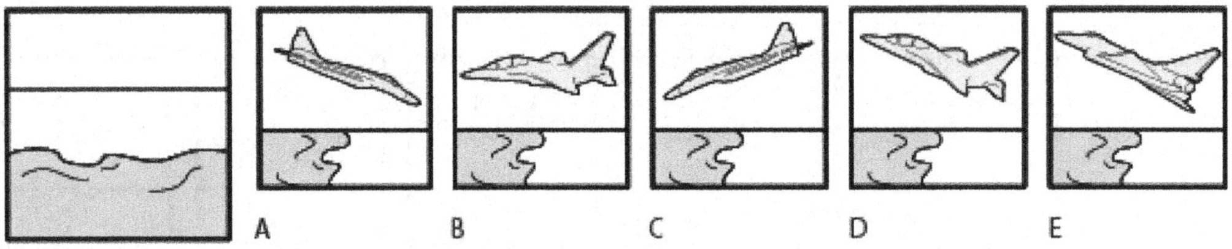

19.

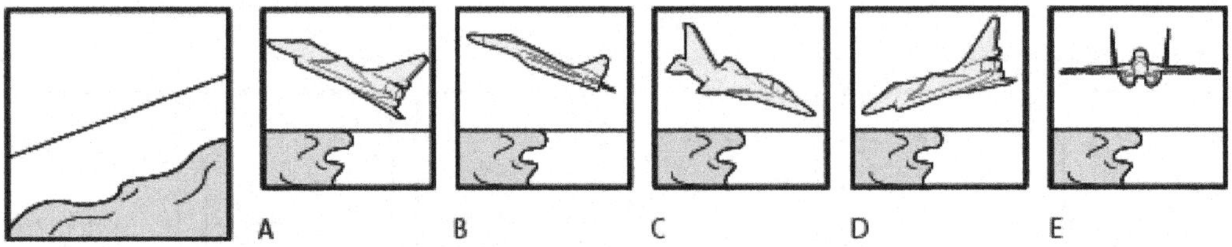

20.

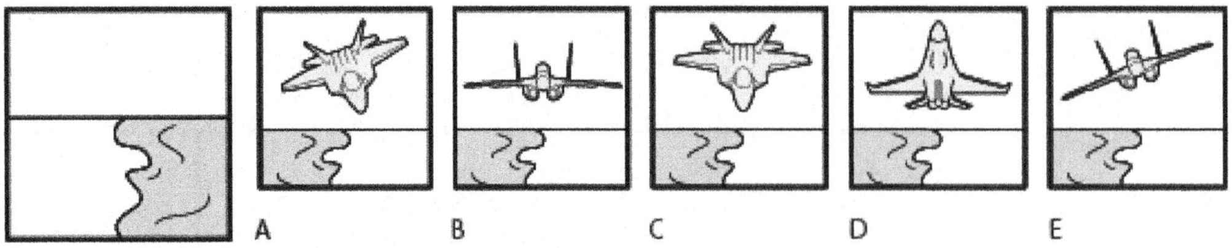

21.

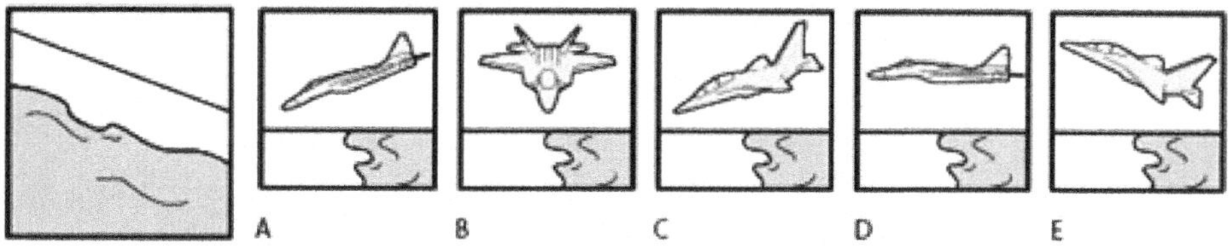

A B C D E

22.

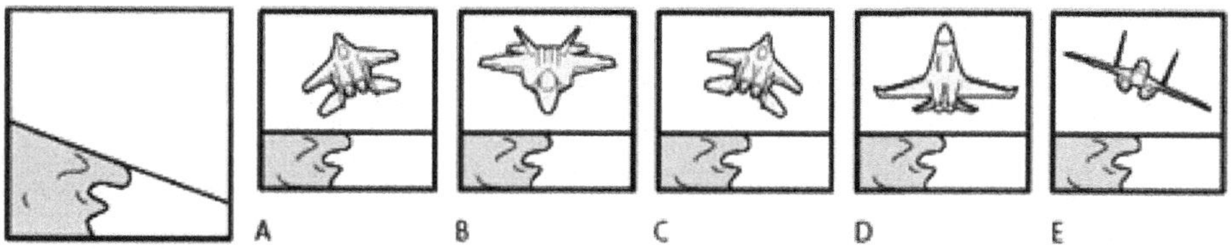

A B C D E

23.

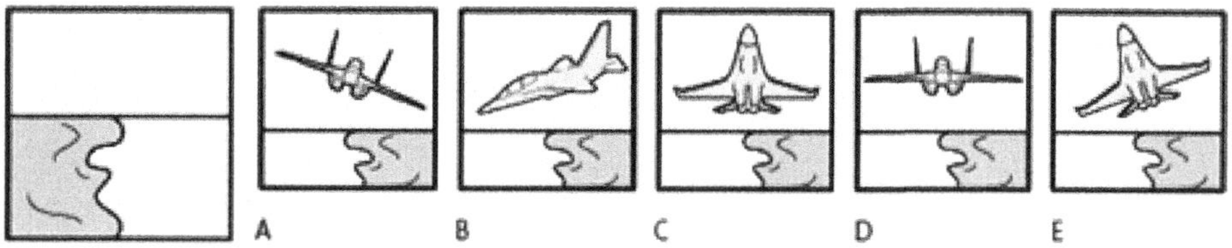

A B C D E

24.

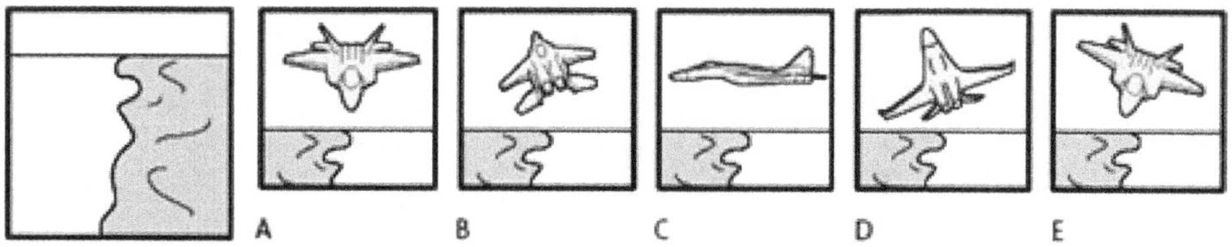

A B C D E

25.

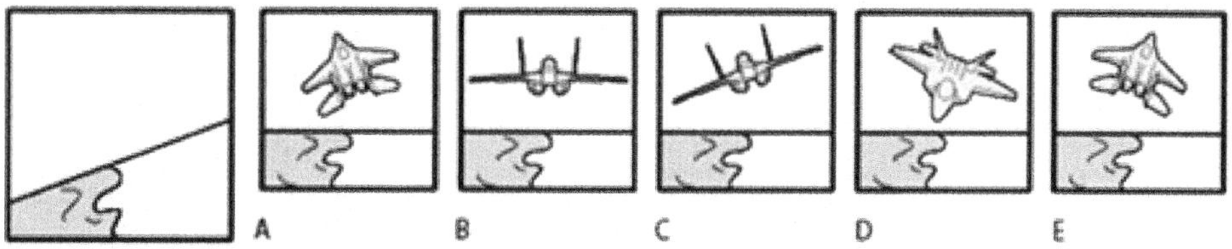

A B C D E

Answers

1. B

2. C

3. E

4. D

5. D

6. B

7. D

8. C

9. B

10. A

11. D

12. C

13. E

14. B

15. B

16. D

17. B

18. A

19. C

20. B

21. C

22. A

23. D

24. A

25. E

Reading Comprehension

Good reading comprehension skills are vital for successful standardized test taking to become a military aviator. Reading comprehension is tested to determine whether prospective aviators are able to parse out critical and accurate information quickly and efficiently. The topics addressed in this section will help in this pursuit.

Topics and Main Ideas

One of the principle skills for effective reading comprehension is the ability to accurately and quickly identify an author's points and intentions. When presented with blocks of written information, it's easy for test takers to feel overwhelmed. One of the best ways to get a handle on a text is to visually divide the writing into manageable chunks.

First, test takers should begin by looking for the author's topic(s) and main idea(s), as there is a difference between them.

A *topic* is the subject of a written selection; it supports an overall theme, idea, or purpose of a passage. Here's where reading comprehension can get tricky. An author can choose to have more than one topic in a written passage, with each building on the previous one to support an overall conclusion or thesis.

The *main idea* is information that expands upon and supports a topic. Main ideas can be thought of as the supporting frame that makes up a building. As a whole, the structure—the building itself—is made up of various sections. Each section is positioned to support other sections and each section contributes to the whole structure. A writing passage offers a topic (the building) that's supported by the main ideas (an internal structure).

Another way to view topics and their main ideas is to consider these three words: *what, how,* and *why.* A written passage has a purpose—a *what*. The topic of a written passage is the *what.* The main ideas expand on *how* and *why*.

It's important to separate topics and main ideas from the rest of the text. In most cases, simply skimming a text selection is effective. Generally speaking, an author will use titles or important first paragraph sentences (called *topic sentences*) to indicate what the passage is about and what each main idea will address.

Review the passage below. In this case, words and phrases that indicate the *topic* appear in **bold**. Words and phrases that indicate *main ideas* appear in *italics*.

> **Forces acting on an aircraft are called the aerodynamics of flight.** *There are four: thrust, drag, lift, and weight.* It's essential for pilots to understand these forces (or aerodynamics) in order to control them during flight. *Thrust* is a forward-moving force caused by power. In the case of aircraft, this power is provided by propellers, rotors, and engines. Thrust works against the force of drag, which is a rear-moving force. *Drag* is caused by air disturbance created by the plane itself in terms of its wings or other protrusions. *Weight* is a downward force caused by gravity and includes factors such as the combined weight of the aircraft and crew. *Lift* is diametrically opposed to weight, acting in a perpendicular motion to the flight path. It's also produced by air effect.

The topic of the passage is the aerodynamics of flight—specifically, the forces acting on an aircraft. The topic is supported by the building blocks—its main ideas—that explain what and how aerodynamic forces affect aircraft. These main ideas include thrust, drag, lift, and weight. The passage continues to explain what each force is by definition and how it acts on an aircraft. By quickly separating the topic and main ideas, it becomes clear what the passage is about and where the author is going. In this case, additional words and phrases have been italicized, as they indicate main ideas. But boiled down, the first sentence and four words (thrust, drag, lift, and weight) indicate the author's topic and main ideas.

Supporting Details

There are many different construction materials in a single building. Supporting details are like those materials. Like steel, bricks, wood, and concrete, the supporting details support the larger structure (the main idea or ideas) and, in turn, the entire building itself (the topic).

Supporting details usually (but don't always) appear after the first sentence in a paragraph. Supporting details always give the specifics to develop and support the main idea. They may describe, explain, catalogue items, identify, or expand with specifics. Understanding supporting details requires careful attention. The following paragraph has its main supporting details highlighted in **bold**. Notice that even the supporting detail of wing curvature is supported with examples.

> Lift is one of the four forces that affect flight and aerodynamics. These forces cause an object to move up, down, faster, or slower. **Lift is affected by the shape of a plane's wings and their curvature.** When air pressure is diminished over the top of the wings, an aircraft will follow an upward movement. **Changing the curvature of an object's wings is a way to manipulate lift. Even everyday objects are affected by potential lift in flight.** A kite, for example, is able to lift due to its curved shape.

Another way to identify supporting details is looking for words such as *first*, *second*, *next*, and *last*, and phrases such as *for example* and *for instance*. These words can indicate that supporting details follow. The paragraph below makes use of these words and phrases:

> Thrust, drag, weight, and lift are four forces that act in aerodynamics. *First*, thrust, or the forward force produced by power, overcomes drag. *Second*, drag is the force that disrupts aircraft in some manner. It's a rearward force that disrupts airflow around distended objects. *Next*, weight is the downward force caused by gravity. *For example*, the combined weight of any crew, baggage, and the craft itself will affect downward pull. *Lastly*, lift is the force that opposes weight, acting perpendicularly to the aircraft's flight path. All forces combined are referred to as the aerodynamics of flight.

In these examples, the first sentence of the paragraph indicates the topic (the four forces of aerodynamics). The subsequent sentences contain supporting details that expand on the topic.

Topic and Summary Sentences

Notice the last sentence in the last example: "All forces combined are referred to as the aerodynamics of flight." This summarizes the main topic and concludes the paragraph. It also re-emphasizes the topic sentence. Topic and summary sentences are another quick way to skim for the topic(s) the author intended and glean information about supporting detail.

Often, a paragraph will not be so neatly structured. Not all passages contain topic and summary sentences. In fact, larger passages of text will only contain transition ideas between supporting details or other topics. Writing structure is frequently complicated and not always conveniently organized within single paragraphs. In these instances, it's necessary to read more in order to identify a topic.

Test takers can think of topic and summary sentences as clear identifiers with consistent characteristics. Both are usually free of new terminology and avoid introducing new ideas. Ideally, a topic sentence appears first and a summary sentence appears last. Supporting details usually appear between them. The previous passage appears again below, now with the topic and summary sentences in **bold**.

> **Thrust, drag, weight, and lift are four forces that act in aerodynamics.** First, thrust, or the forward force produced by power, overcomes drag. Second, drag is the force that disrupts aircraft in some manner. It's a rearward force that disrupts airflow around distended objects. Next, weight is the downward force caused by gravity. For example, the combined weight of any crew, baggage, and the craft itself will affect downward pull. Lastly, lift is the force that opposes weight, acting perpendicularly to the aircraft's flight path. **All forces combined are referred to as the aerodynamics of flight.**

That was a fairly straightforward illustration, but what if a passage isn't so clearly constructed? Consider this example:

> Crew, baggage, and the aircraft itself all affect the force of weight. Lift becomes important in counteracting weight, acting perpendicularly to an aircraft's flight path. A thrusting force is produced by power and counteracts drag. Drag is the force that disrupts a craft's airflow. All these forces combined act within aerodynamics.

In the above passage, there's no "first" topic sentence. Simply assuming the last sentence is the summary sentence doesn't give all the information. At best, the last sentence indicates the paragraph is about combined forces acting within aerodynamics. Therefore, it's necessary to read the entire paragraph. Parsing out words and information becomes critical. By looking at the first words of each sentence, including potentially unfamiliar words, and at words separated by punctuation like commas, the main topic becomes easier to identify:

> *Crew*, *baggage*, and the *aircraft* itself all affect the *weight* of *force*. *Lift* becomes important in counteracting weight, acting perpendicularly to an aircraft's flight path. A *thrusting force* is produced by power and counteracts drag. *Drag* is the force that disrupts a craft's airflow. All these forces combined act within aerodynamics.

Paragraphs without concise topic and summary sentences can usually be understood with more detailed attention to reading.

Predictions Based on Prior Knowledge

Guessing what will happen next in a written passage is based on prior knowledge. This is called a *prediction*. Prior knowledge is the information a reader already possesses. This skill is important in reading comprehension, as it allows the reader to logically predict what's ahead based on previous reading and experience. Predict what happens next in the below passage:

> The aircraft began rocking and rolling without warning. The plane rolled to the right, then to the left. There was a definite change in pressure. A loud metallic sound screeched throughout the cabin. Things were definitely not going according to plan any longer.

Based on the passage, the reader knows the aircraft is in some sort of distress. Furthermore, based on prior knowledge, the reader may predict one of two things: the aircraft will land safely or it won't. Notice that predicting the outcome while reading can be very tricky. Predictions are not always accurate (one way or another); but in the passage above, only two predictions are possible. These predictions are based on a reader's prior knowledge of flight experience and common sense. Determining the root cause of the aircraft's issue and final outcome would require further information. Therefore, a reader must be willing to readjust predictions as new details become available.

Making Inferences

An *inference* is a conclusion based on evidence and reason. Making inferences involves determining information that is implied rather than stated outright. A reader cannot usually infer conclusions merely by reading the first and last sentences of a passage. Making inferences requires attention to detail. It's important not to go beyond the written information in a passage and come to conclusions that are not inferred by the author. This can be tricky, but practice will help. Consider the following sentence:

> The pilot was relieved as the airstrip suddenly appeared below.

From this sentence, the reader can infer that the pilot was having difficulty during the flight. But one cannot infer that bad weather or an incorrect flight path was the cause of the problem. The reader also cannot infer that the complication was even resolved or that the flight ended safely. The reader can only infer that something unexpected occurred during the flight and, as a result, the pilot was relieved to make visual contact with the landing area. Consider more information below:

> The pilot was relieved as the airstrip suddenly appeared below. It had been a long journey. His head pounded and sweat poured down his back. His breathing was labored, and his chest felt heavy. He was burning up.

At this point, the reader can make new inferences based on the additional information. One is that the pilot felt ill. Another is that a problem caused excessive heat in the aircraft. Still, the reader cannot come to a final conclusion based on the passage. Although the sentences above provide detail about the pilot's condition, the reader can only make an educated guess as to the actual problem. Finally, read the passage that follows:

> The pilot was relieved as the airstrip suddenly appeared below. It had been a long journey. His head pounded and sweat poured down his back. His breathing was labored and his chest felt heavy. He was burning up. His nerves were frayed, and he was at wits' end. He never thought the flight simulation would be so difficult to navigate, but he'd done it, and he was about to bring it home.

At this point, the reader has enough detail to infer that the pilot was in training and simply glad to be on his way to finishing a difficult flight simulation. Keep in mind that making an inference involves drawing a conclusion based on evidence and on reason. The evidence is the pilot's demeanor, physical symptoms, and, most importantly, the words *flight simulation*. Reason dictates that this is most likely a training situation. Inference here, however, doesn't indicate that the pilot successfully completed the simulation in good health.

Drawing Conclusions

Using inference techniques and their own life experience, readers should be able to draw conclusions from the information they read. Words such as *may*, *can*, and *often* indicate calls for a conclusion. Again, readers must be careful to draw conclusions based only on the information contained in a particular passage. They should not embellish or fill in the blanks with their imagination. Making faulty assumptions can lead to incorrect conclusions and the wrong test answers.

In a reading comprehension passage, test takers should look for key words indicating that drawing a conclusion is appropriate, then they should check the corresponding test answers. It is generally recommended to avoid answers that use all-or-nothing indicators. Words and phrases such as *always* and *never* can usually be eliminated.

Sequence

The order within text is called *sequence*. Test takers should look for identifying words such as *initially*, *then*, and *last,* as well as words that indicate steps such as *first*, *second*, and *next*. Reading a passage for details that aren't in the correct order will lead to incorrect test answers, especially if the text covers specific how-to information.

Sequence may also be implied. Read the sentence below:

> The pilot prepared the flight plan, double-checked her gear, and then boarded the plane.

Clearly, the pilot boarded the plane last. She didn't double check her gear first, then prepare the flight plan. While the sentence doesn't state the pilot's actions in numerical steps, the sequence is implied by the position of the phrases and logical inference.

Sequence doesn't always occur first to last. Sometimes, an author may reverse sequence or imply sequence in a way that may require readers to keep track. In this instance, making a brief outline can be helpful.

Let's practice sequence. Read the following paragraph:

> Obtaining a reliable weather report before taking off is an essential part of flight. First, have the flight route clearly mapped. Next, prepare any information that may be required by organizations that provide weather information (for example, pilot qualifications and the type of flight planned). Then, check any and all sources available to insure the information is consistent and accurate.

In the above passage, sequence is clearly implied by the use of the words *first*, *next*, and *then*. The steps follow in order. A pilot must determine a flight route before checking weather reports. Additionally, deciding if the flight will follow visual flight rules or instrument flight rules (the type of flight) ahead of time will affect the weather information the pilot needs to consider. Conducting these steps out of sequence simply makes no sense and could lead to a risky flight.

Comparison and Contrast

A reader must be able to compare and contrast information within a text in order to comprehend an author's meaning. *Comparison* involves relating two or more concepts or objects by finding commonalities. *Contrasting* is finding dissimilar characteristics between concepts or objects. Think of the construction materials referred to earlier: bricks, mortar, wood, concrete. Comparing these materials involves determining what they have in common. All are materials used in construction. All are available for builders to use according to a construction plan, and all of these materials are used widely throughout the industry. Contrasting them involves determining how they are different. Brick is certainly made differently from concrete and uses different materials. Wood is organic. Mortar is the material that holds brick together. Being able to compare and contrast allows the reader to link ideas and to differentiate between concepts.

Cause and Effect

Text may be structured to show cause and effect. *Cause* is an event that results in an occurrence. *Effect* is the direct result or results of the cause. In order to identify cause and effect relationships, readers should look for words such as *because, since, due to*, or *as a result*. Text that begins with *consequently* or *therefore* also indicates cause and effect.

Cause and effect might not be directly stated. In the sentence, *"The pilot neglected to consult all available weather information, and the flight didn't go as planned,"* a direct cause and effect relationship is implied. Because the pilot didn't consult all available weather information (the cause), the flight didn't go as planned (the effect). In this example, no direct words indicate cause and effect; however, cause and effect are directly related.

Identifying an Author's Position

Many times, standard reading comprehension assessments will require the reader to be able to identify an *author's position*. This is the stance or belief the writer states or implies. When considering a writer's overall message, it's important for readers to be on the lookout for position. Even factual text can take on bias. An author's position can be clearly stated, such as in a passage that argues an opinion, or it can be implied, such as a text that uses emotional language without a definitive statement of belief.

Identifying an Author's Purpose

An author's purpose is closely aligned with an author's position; however, the two are different. An *author's purpose* is the reason for the text itself. The purpose of the text may be to entertain or inform. An author may try to convince the reader through a firm statement and subsequent data to back up an argument. A persuasive purpose should be approached with caution, as the author clearly has an agenda and wishes to persuade the reader into agreement.

Authors may approach purpose in a variety of ways. The stronger the emotional language and the more information an author presents to argue a particular position, the more persuasive the intent. When a passage doesn't take a particular stance, but instead, is primarily telling a story, its purpose is more likely to entertain.

Readers should be vigilant for text that makes claims. Such text is persuasive. Text that gives information without making claims is likely informative in nature. The purpose may be to instruct the reader or to help the reader reach factual conclusions.

Word Meaning From Context

Stumbling over unfamiliar vocabulary is common; therefore, a reader must be able to identify word definitions from context. Determining *word meaning from context* involves using the words around an unfamiliar term to determine its meaning. Read the sentence below and pay attention to the word in italics:

> The pilot had been *apprised* of the weather well in advance; however, all the information she'd been provided still didn't quiet her fears.

In this instance, the reader may be unfamiliar with the word *apprised*. By looking at the word within its context, it's likely that *apprised* means informed. The first phrase indicates the pilot has been given weather information in advance. The second phrase actually states she'd been provided with that information before the flight.

Sometimes, a definition may not be so clearly identifiable. Read the sentence below and pay attention to the word in italics:

> A *horde* rushed towards *The Spirit of St Louis* that day, but off to the side, all alone on the Le Bourget field, stood a small boy who dreamed of one day becoming a pilot like Charles Lindbergh.

This sentence requires the reader to infer the meaning of the word *horde* using comparison and contrast. Because the passage contrasts people rushing to the plane with a lone boy standing off to the side, the reader must infer that *horde* means a large group.

Identifying a Logical Conclusion

Being able to identify a logical conclusion is an essential skill in reading comprehension. *Identifying a logical conclusion* is being able to form an opinion after reading a textual passage. This can help readers determine if they agree or disagree with an author. Clarifying a logical conclusion requires a reader to keep track of all pertinent points in a written passage. It also requires the reader to ask questions while reading and searching for answers that an author may provide. It's possible to draw several conclusions from a particular passage. An author may not provide a clearly stated conclusion, so readers should be careful. Logical conclusions should be directly supported by text. Readers should avoid "reading into" a passage and inventing supported text that doesn't exist in order to come to a conclusion.

Practice Test

Reading Comprehension Test

Aircraft propulsion must achieve two objectives: it has to balance the thrust and drag of a craft in cruising flight, and thrust must exceed drag in order to accelerate movement. Balancing the thrust and drag of a craft in cruising mode follows Sir Isaac Newton's first law of motion. This law states that an object in motion will remain so, in a straight line, with the absence of a net force to affect a change. In other words, if all external forces cancel each other, an object remains at a constant velocity. This is necessary to understand the idea of flight at a constant altitude. However, if a pilot affects a change to a craft's thrust, the balance between that thrust and the drag of a craft is disrupted. The net forces have been changed in a manner that will affect acceleration and propel the aircraft until there is a new balance between thrust and drag in velocity.

1. What is the topic of the paragraph above?
   a. Aircraft propulsion is a difficult concept to understand.
   b. External forces must cancel each other out in order for an aircraft to fly.
   c. An aircraft that remains at a constant velocity will remain at a constant altitude as well.
   d. Aircraft propulsion involves two objectives involving the laws of motion.
   e. Thrust must exceed drag in order for an aircraft to accelerate.

2. Choose the option below that does NOT support the topic of this paragraph:

   One of the four forces that affects aerodynamics is weight. The total weight of an aircraft must be factored into the mechanics of flight. Lift also affects flight and must be considered. Weight must include the craft itself, accompanying gear, the fuel load, and any personnel. Carefully assessing weight will help a pilot calculate how to oppose the downward force of weight during flight.

   a. The sum weight of an aircraft is important to consider.
   b. Personnel should be calculated in overall weight.
   c. The mechanics of flight also includes lift.
   d. Weight is a downward force in aerodynamics.
   e. Fuel load cannot be eliminated in measuring weight.

3. Read the following paragraph, then select the topic:

   Standard weather briefings can be obtained from a variety of sources, including the FAA. Pilots can obtain abbreviated weather information from briefers to supplement available electronic information. Information that a pilot has previously obtained should be provided to the briefer in case conditions have changed. An outlook weather briefing is obtained six or more hours in advance of the flight. Lastly, it's important to obtain in-flight briefings to monitor ongoing weather conditions during the flight course itself.

   a. In-flight briefings monitor ongoing weather conditions during flight.
   b. Briefers are people who provide meteorological information to pilots.
   c. Standard weather briefings are the most comprehensive types.
   d. All briefings should be updated.
   e. There are a variety of weather briefing types.

4. Which of the options can be inferred by the passage below?

Don't attempt to fly through thunderstorms. Try to safely and wisely circumnavigate them. Know your personal skills and limitations regarding weather flight before you gather forecast information. Keep in mind that a variety of conditions in-flight can change quickly. Don't fly into areas of rain where the air temperature is at or close to freezing. Always allow more margin of error for weather at night, and do not attempt to navigate through cloud "holes."

a. Flying in inclement weather involves using sound judgment.
b. Thunderstorms are not navigable.
c. Changing conditions will lead to flight trouble.
d. Freezing rain can be managed if the pilot knows what he is doing.
e. Flying at night is tricky.

5. Read the following passage and select the best definition for the word in **bold**:

Everyone had said his piloting **acumen** was not to be questioned. He had a knack for feeling the aircraft, as if it was an instinctual part of himself. He had a knowledge base he'd built up over years of study and thousands of flying hours; but moreover, he had the ability to think and act quickly in the most unexpected circumstances.

The word *acumen* means:

a. Ability to fly
b. Knowledge
c. Shrewdness and keen insight
d. Visual ability
e. Impressive number of flight hours

6. Read the following passage and select the best conclusion:

Piloting is a consummate skill to obtain, but it isn't an impossible goal. A potential pilot must be versed in flight instruction, which consists of many classroom hours and even more flight hours. A potential pilot needs to have a logical mind, be armed with the right information for each flight, and be correctly trained in all piloting procedures. Anyone who lacks a crucial ingredient in training takes a serious risk when flying. With the right education and the right instruction, becoming a pilot is a rewarding venture.

a. Anyone can become a pilot.
b. Becoming a pilot is a risky proposition.
c. Innate ability is key in flight instruction.
d. Learning to fly, if taught and practiced correctly, is a fulfilling skill.
e. Piloting can be impossible to learn.

7. Read the following passage and identify the statement that best describes the author's position.

> Piloting is a consummate skill to obtain, and for many, it's a skill that remains unattainable. A potential pilot must be so well-versed in flight instruction that studying the requirements can seem overwhelming. In addition, thousands of flight hours must be logged in practice before a pilot can begin a career. The physical and mental demands made on many potential students can be prohibitive. It's a dedicated person who perseveres and becomes successful in the flight industry.

a. Piloting is a difficult goal requiring serious consideration.
b. Becoming a pilot is cost-prohibitive.
c. Piloting careers are few and far between.
d. The physical and mental demands on piloting students should be the main consideration when deciding on a career in flight.
e. Piloting involves a lot of luck and practice.

8. Carefully read the following passage, taking note of the letters that precede each sentence. Then select the option that best describes the text's correct sequence:

> Aviation's history expands over 2,000 years. A) Eventually, Leonardo da Vinci studied the flight of birds and began noting some basic principles of aerodynamics. B) It is said that kites were the first airborne, manmade "vehicles." They existed in China around the fifth century B.C. and were capable of carrying a person into the air. C) The first recorded attempt at flight was made by a Spanish citizen who covered his arms in feathers and jumped. D) The late 18th century saw a major development in ballooning, and the steerable airship, or dirigible, was developed. E) Sir George Cayley significantly impacted aerodynamics in the early 19th century by contributing key ideas to flight physics. F) In 1903, Orville Wright flew 130 feet in twelve seconds.

a. The sequence is correct
b. B, D, C, A, F, E
c. A, C, F, D, E, B
d. C, B, A, D, E, F
e. D, F, E, B, C, A

9. Which of the following best states the author's purpose?

> While many innovative aircraft have been developed throughout the history of flight, certain aircraft can be referred to as groundbreaking and revolutionary. Their existence has transformed flight and advanced the study of aviation. The Wright Brothers' 1905 plane, The Flyer, was the first to have three-axis control. It had innovative seating for its pilot and attained faster speeds due to increased power. The Monocoque, developed by Deperdussin manufacturers, introduced a new shell structure capable of handling the forces of stress. In 1919, Hugo Junkers introduced all-metal structures with thicker wings. The Douglas DC-1 bears the title of the first American, scientifically-designed structure.

a. The purpose is to inform by presenting examples of revolutionary aircraft inventions throughout flight history.
b. The purpose is to entertain the reader with stories about airplanes.
c. The purpose is to argue that Deperdussin's model, which handled the forces of stress, was the most revolutionary development in flight.
d. The purpose is to convince the reader that American, scientifically-designed structures were the best.
e. The purpose is to convince the reader that these examples are the only ones worth considering in the development of innovative flight.

10. Read the following passage, then select the option that best identifies cause and effect:

> While it can be dangerous, turbulence during flight is common and manageable. Turbulence is a disruption in air flow. Think of it as a bump on the road while driving to work. It happens all the time and, when managed effectively, it's fairly benign. One type of turbulence is CAT (Clear Air Turbulence). It cannot be seen, and one way to avoid it is to fly at a higher, and likely smoother, altitude. Severe turbulence, a disturbance that results in an altitude deviation of 100 feet, is uncomfortable, but not dangerous.

a. Turbulence causes plane crashes.
b. Turbulence is caused by a disruption in air flow.
c. Turbulence is common and the effects are always dangerous.
d. Turbulence is uncommon and the effects are usually benign.
e. Turbulence is manageable by flying at a higher altitude.

11. Read the same, expanded passage below, then select the answer that best states the outcome:

> While it can be dangerous, turbulence during flight is common and manageable. Turbulence is a disruption in air flow. Think of it as a bump on the road while driving to work. It happens all the time and, when managed effectively, it's fairly benign. One type of turbulence is CAT (Clear Air Turbulence). It cannot be seen, and one way to avoid it is to fly at a higher, and likely smoother altitude. Severe turbulence, a disturbance that results in an altitude deviation of 100 feet, is uncomfortable, but not dangerous. A disruptive air flow that causes turbulence doesn't have to result in difficulty for the pilot. A well-trained pilot and crew can manage turbulence so that it isn't disruptive to passengers in flight.

a. If not managed, turbulence will result in disaster.
b. Weather maps predict CAT.
c. Flying at a higher altitude always solves the problem of turbulence.
d. A bump in the road is potentially more invasive than air turbulence.
e. Encountering turbulence is not a significant issue for pilots and will not likely result in air disaster.

12. Read the passage below and select the option that best states the author's purpose:

> The mechanics of flight are not difficult to understand, although the concepts often contain difficult terminology. Words such as yaw, pitch, and roll define aircraft axes, or dimensional center lines of gravity that intersect an airplane. Understanding these terms is crucial to understanding the mechanics of flight. Controls such as actuators generate forces that act on an aircraft. Rotational dynamics operate in tandem with translational ones to affect an aircraft's position and trajectory.

a. The purpose of the passage is to convince the reader that the mechanics of flight are easy to understand.
b. The purpose of the passage is to confuse the reader by using difficult terminology to explain a simple concept.
c. The purpose of the passage is to inform the reader about terminology in the mechanics of flight.
d. The purpose of the passage is to describe how axes and aircraft controls affect the mechanics of flight.
e. C and D.

13. Read the below paragraph and select the option that likely defines the word in **bold** within context:

> A helicopter pilot maneuvers many controls to maintain **equilibrium** in flight. These controls include the cyclic stick, anti-torque pedals, and the collective lever. The cyclic control maintains the pitch angle of the craft's rotating blades. The anti-torque pedals control the nose of the helicopter. The collective lever controls the pitch angle of the rotating blades together or independently as needed. All three controls help the pilot stay stable in flight.

a. Equilibrium means unsteady.
b. Equilibrium means pitchy.
c. Equilibrium means the collective pitch angle of a helicopter's blades.
d. Equilibrium means flight stability.
e. Equilibrium means the controls used to fly a helicopter.

14. Read the paragraph and select the option that best states cause and effect:

> The pilot looked to the left, and there it was: a tiny glimmer of white on the left wing. He was on the leading edge of it. He knew the temperature had dropped. He'd checked it repeatedly. He had been cleared at 9,000 feet but needed to go down. He needed to get to warmer air and get there quickly. He checked the terrain clearance and decided it was now or never. He popped out of the clouds and began his descent, chunks of ice flying off the fuselage.

a. Not checking a weather report ahead of time led to ice on the aircraft.
b. Flying too high resulted in icing.
c. The pilot didn't have the right clearance and encountered icing as a result.
d. A sudden, unexpected drop in temperature resulted in icing that the pilot had to account for.
e. Icing led to additional weight on the fuselage.

15. Read the paragraph and select the option that best defines the words in **bold**:

> Differential pitch control is a factor in helicopter flight. Differential pitch control affects the yaw axis and causes the helicopter to spin in the direction of its rotational, tilted rotors. **Coaxial** mounted helicopters, or crafts with dual rotors mounted above each other concentrically, require interaction between rotors. **Intermeshing** rotors, or helicopter rotors at a set angle to each other and turn in opposite directions, require increasing collective pitch on one blade set.

a. From the passage, it's impossible to tell what coaxial and intermeshing rotators are.
b. Coaxial rotors are dual and mounted above each other, while intermeshing rotors are angled apart from each other.
c. Coaxial means interaction and intermeshing means turning in an opposite direction.
d. Coaxial and intermeshing mean the same thing.
e. Coaxial and intermeshing rotors have nothing to do with differential pitch control.

16. Read the passage below and select the option that best identifies the main idea:

> The synchropter arrangement was first designed for use in Germany's anti-submarine war craft. Often referred to as intermeshing rotor configuration, this design sets two oppositional rotors at a slight angle from each other in a transverse manner so that they actually intermesh. The configuration allows a helicopter to function without a tail rotor in order to save power.

a. The development of the synchropter began during World War II.
b. Synchropter configuration is best understood as an intermeshing rotor configuration, oppositional in nature.
c. Oppositional rotors always intermesh.
d. Functioning without a tail rotor saves power for a helicopter.
e. Transverse rotors always intermesh.

17. Choose the option that best defines the word in **bold** below:

A helicopter pilot's use of control in a hover is critical to success. Correct use of the cyclic, the collective, and the anti-torque pedal controls help maintain balance. The cyclic is used to cut down on **drift** in the craft's horizontal plane. The collective helps maintain altitude, and the anti-torque control is used to direct the craft's nose direction.

a. Drift means unwanted movement within the helicopter's horizontal plane.
b. Drift refers to stable air around the helicopter.
c. Drift refers to the pilot's ability to control the cyclic.
d. Drift means hover.
e. Drift can be defined as unwanted movement within the helicopter's vertical plane.

18. Read the passage and select the answer that best states the logical conclusion:

A chain of events often defines incidents that lead to an accident. In aviation, this chain of events is often referred to as the "error chain." This term purports that it isn't one single event that leads to an air disaster but a chain of separate events. In other words, an accident is caused by more than one problem and most likely has more than one root cause. These events can be caused by human error or mechanical issues. A recent study indicates that pilot error is more likely to contribute to air disasters than mechanical problems.

a. Air disaster is unavoidable.
b. The error chain is specific to aviation.
c. A pilot's appropriate risk assessment before and during a flight will cut down on negative error chain events.
d. Mechanics aren't a consideration in the error chain.
e. Mechanics should be the primary consideration during flight.

19. Read the passage and select the correct answer to the question below:

Proper risk assessment for flight is critical for avoiding air disaster. As most events in the error chain can be addressed and controlled by the pilot, it's essential to consider all risks before and during flight. A pilot must perform all necessary informational checks prior to a scheduled flight. The flight path itself must be registered and approved. During the flight, a pilot must communicate regularly with others in the air and on the ground. Making sudden, unapproved changes during flight can lead to a negative chain of events. Being unprepared, uncommunicative, and unaware are all mistakes a pilot can make while on the job. Constant risk assessment is necessary to a successful and happily uneventful flight.

Which of the following statements provides the most encompassing overview with which the author would most likely agree?

a. Risk assessment includes mechanical checks and balances.
b. The pilot is the leading force in appropriate risk assessment.
c. Risk assessment is best conducted on the ground prior to flight.
d. Unapproved changes during a flight will always lead to problems.
e. Communication is the key factor in a successful flight.

20. Read the passage below and select the best meaning for the word in **bold**:

The United States attack helicopter came into its own during the 1990s. During Operation Desert Storm, the Apache fired against early warning radar sites to, in part, support ground troops. In addition, Apaches were able to destroy surface-to-air (SAM) sites via their Hellfire missile technology. The missiles **obliterated** many armored cars and tanks as well.

a. Obliterated can be defined as a type of Apache missile.
b. Obliterated means thoroughly destroyed.
c. Obliterated is defined as a military strategy.
d. Obliterated means flying low over armored cars and tanks.
e. Obliterated can be defined as an Apache flight technique, specific to Operation Desert Storm.

Answer Explanations

1. D: Choice *A* can be eliminated, as this idea isn't stated in the selection. Choice *B* only represents a portion of the subject. Choice *C* is stated in the passage, but the paragraph isn't solely about constant velocity. Choice *E* is a supporting concept, not the main topic. Only Choice *D* answers the question.

2. C: Choices *A*, *B*, and *E* are stated in the passage. The sum weight of an aircraft is important to calculate, as are personnel and fuel load. Notice that Choice *E* states fuel load *cannot* be eliminated in measuring weight. Careful comprehension is important when reading Choice *E*. Choice *D* also supports the idea that weight affects aerodynamics. In Choice *C*, while it's true that the mechanics of flight includes lift, that detail doesn't support the overall topic of weight.

3. E: Choices *A*, *B*, and *D* are true, but they don't encompass all of the ideas presented in the passage. Choice *C* isn't stated and is incorrect. Choice *E* encapsulates all of the ideas presented in the passage.

4. A: From the provided paragraph, a reader may infer that a pilot should use sound judgment when flying. Choice *B* is not stated. Choice *C* is stated as fact but isn't true. Choices *D* and *E* are false as stated.

5. C: All of the options indicate some aspect of the pilot's acumen, but each is a contextual clue that, when put together, indicate the pilot has an overall shrewdness and keen insight when it comes to his flight talent. Choice *A* is incorrect as *acumen* isn't specific to flight. Choice *B* is insufficient, for knowledge alone doesn't encompass the other passage descriptors. The same can be said for Choice *D*; and, while Choice *E* is directly mentioned in the paragraph, an impressive number of flight hours is just one accomplishment of the pilot and too specific to encompass the pilot's other skills.

6. D: Choice *A* is incorrect; the passage doesn't state that anyone can become a pilot. The text implies that becoming a pilot carries some risk, but Choice *B* is not the correct answer, as the rest of the paragraph doesn't support that statement. Choices *C* and *E* are incorrect because they aren't stated. Only Choice *D* encompasses all of the text as the logical conclusion.

7. A: While the reader may disagree with the author's position, all of the statements in the passage indicate the author's position is that piloting is a difficult goal that requires serious consideration on the part of a potential flight student. Choices *B*, *C*, and *E* aren't stated. Choice *D* can be ruled out because the physical and mental toll on piloting students isn't the main position of the paragraph.

8. D: The text indicates a sequence throughout time. Sentence C relates the very first recorded attempt at flight. Sentence B indicates the first manmade vehicles in fifth century China. Sentence A addresses Leonardo da Vinci's famous work in flight during the Renaissance. Then, Sentences D, E, and F occur in the correct historical order.

9. A: The purpose is to inform the reader of certain facts by presenting examples of revolutionary aircraft throughout flight history. Choice *B* is incorrect. While the passage may be fun to read, its primary purpose is not entertainment. It doesn't tell a story. Choice *C* is incorrect. The main focus of the passage is not Deperdussin's model. Choice *D* is incorrect. The passage doesn't indicate that American structures were better than any other. Choice *E* is also incorrect. The reader would have to assume that the author intended to argue that these examples of aircraft development are the only ones worth considering. The first paragraph of the passage indicates they aren't.

10. B: Turbulence is caused by a disruption in air flow. Choices *A*, *C*, and *D* are incorrect as stated. Choice *E* is true, but it doesn't indicate a cause and effect relationship.

11. E: Choices *A*, *B*, and *D* are incorrect as stated. Choice *C* is likely correct but it isn't always correct as stated, and it doesn't encompass a conclusion or outcome. Only Choice *E* summarizes the best potential outcome to encountering turbulence.

12. E: The answer is Choice *E*, which encompasses the idea that the author's purposes are both to inform and describe. Choice *A* doesn't address the entire purpose. Yes, the author states the mechanics of flight are not difficult to understand, but that's not all the author does in the passage. Choice *B* is incorrect as stated.

13. D: Notice that the main idea of the passage—maintaining equilibrium in flight—is stated in the first sentence and re-stated in the last. The word *equilibrium* is re-defined for the reader as staying stable in flight. All other options are incorrect as stated.

14. D: The reader isn't given enough information to determine if Choice *A* is correct. Nowhere in the passage does the author indicate that 9,000 feet is too high, so Choice *B* is incorrect. The pilot had the correct clearance, so Choice *C* is incorrect as stated. Although icing led to additional weight on the fuselage, Choice *E* doesn't best address the cause and effect relationship depicted in the passage.

15. B: All other four options are incorrect as stated. Notice that after each bolded word, the author actually defines the words *coaxial* and *intermeshing*, as both are used to describe types of rotor configurations in helicopter design.

16. B: The main idea of the paragraph is best stated in Choice *B*. Choices *A* and *D* are true, but they don't reflect the entire main idea. Choices *C* and *E* are incorrect as stated. Notice that both options use the word *always*. Be wary of test options that use absolutes; rarely are such options correct.

17. A: Although the passage doesn't directly define drift, the reader can use previous knowledge in answering the question correctly. In addition, the reader can eliminate all other options as incorrect because none of them are true.

18. C: Choice *A* is untrue. Choice *B* doesn't state a conclusion applicable to the entire passage. Choice *D* is untrue as stated. Although mechanics should be considered during flight, Choice *E*, as it's written, doesn't state a conclusion applicable to the information in the passage.

19. B: Choice *A* is correct but doesn't encompass all of the main ideas in the passage. Choice *C* is untrue. Choice *D* is partially true but again, note the use of the word "always." In this case, making unapproved changes during flight (while highly inadvisable) doesn't always lead to problems. Choice *E* isn't true as stated. While communication is important to successful flight, it's not the key factor. As the author stated that most events in the error chain can be addressed and controlled by the pilot, Choice *B* is the best answer.

20. B: In this case, all other choices are incorrect. By using context around the word *obliterated*, the reader can infer that the meaning is "thoroughly destroyed."

Math Skills Test

Order of Operations

Numbers usually serve as an adjective representing a quantity of objects. They function as placeholders for a value. Numbers can be better understood by their type and related characteristics.

Integers

An integer is any number that does not have a fractional part. This includes all positive and negative whole numbers and zero. Fractions and decimals—which aren't whole numbers—aren't integers.

Prime Numbers

A *prime* number cannot be divided except by 1 and itself. A prime number has no other factors, which means that no other combination of whole numbers can be multiplied to reach that number. For example, the set of prime numbers between 1 and 27 is {2, 3, 5, 7, 11, 13, 17, 19, 23}.

The number 7 is a prime number because its only factors are 1 and 7. In contrast, 12 isn't a prime number, as it can be divided by other numbers like 2, 3, 4, and 6. Because of they are composed of multiple factors, numbers like 12 are called *composite* numbers. All numbers greater than 1 that aren't prime numbers are composite numbers.

Even and Odd Numbers

An integer is *even* if one of its factors is 2, while those integers without a factor of 2 are *odd*. No numbers except for integers can have either of these labels. For example, 2, 40, -16, and 108 are all even numbers, while -1, 13, 59, and 77 are all odd numbers, since they are integers that cannot be divided by 2 without a remainder. Numbers like 0.4, $\frac{5}{9}$, π, and $\sqrt{7}$ are neither odd nor even because they are not integers.

Decimals

A decimal number is designated by a decimal point which indicates that what follows the point is a value that is less than 1 and is added to the integer number preceding the decimal point. The digit immediately following the decimal point is in the tenths place, the digit following the tenths place is in the hundredths place, and so on.

For example, the decimal number 1.735 has a value greater than 1 but less than 2. The 7 represents seven-tenths of the unit 1 (0.7 or $\frac{7}{10}$); the 3 represents three-hundredths of 1 (0.03 or $\frac{3}{100}$); and the 5 represents five-thousandths of 1 (0.005 or $\frac{5}{1000}$).

Rational and Irrational Numbers

Rational numbers include all numbers that can be expressed as a fraction; in other words, rational numbers encompass all integers and all numbers with terminating or repeating decimals. That is, any rational number either will have a countable number of non-zero digits or will end with an ellipses or a bar (3.6666... or $3.\bar{6}$) to depict repeating decimal digits. Some examples of rational numbers include .25, 12, -3.54, $110.\overline{256}$, $\frac{-35}{10}$, and $4.\overline{7}$.

Irrational numbers include all real numbers that aren't rational. An irrational number can be thought of as any number with endless non-repeating digits to the right of the decimal point. They can be expressed as an endless decimal but never as a fraction. The most common irrational number is π, which has an endless and non-repeating decimal, but there are other well-known irrational numbers like e and $\sqrt{2}$.

Real Numbers

Defined by Descartes in the seventeenth century, real numbers include all numbers found on an infinite number line. All irrational and rational numbers are real numbers. Non-terminating decimal numbers and π are also real numbers. As the range of real numbers extends to both negative and positive infinity, the set of real numbers is complete and uncountable. This set is known as the complete ordered field of numbers.

Roman Numerals

The mathematical legacy of ancient Rome is alive in modern times despite its limited application to counting or customary usage. The Romans expressed their numbers through the use of representative symbols. Their numbering system doesn't account for 0, decimals, or negative numbers.

Arabic Numeral	Roman Numeral
1	I
5	V
10	X
50	L
100	C
500	D
1,000	M

Arabic Numeral	Roman Numeral	Arabic Numeral	Roman Numeral
1	I	6	VI
2	II	7	VII
3	III	8	VIII
4	IV	9	IX
5	V	10	X

There are two important steps to accurately reading and writing in Roman numerals. First, any smaller number immediately preceding a larger number is deducted from the larger number. For example, XC equals 90, as 10 is less than 100, so it's subtracted, as in 100–10=90. Secondly, any smaller numbers after the largest one are added.

Here's an example: XLIII equals 43. In the first step, X is smaller than L, so its value is deducted from that of L, leaving 40. In contrast, the III to the right of the L is added to the remaining 40, giving a result of 43.

Rounding Numbers

It's often convenient to round a number, which means to give an approximate figure to make it easier to compare amounts or perform mental math. When rounding to a certain place value, consider the next digit after that place value. When that digit is 5 or more, the digit in the selected place value gets rounded up. The digit used to determine the rounding, and all subsequent digits, become 0, and the selected place value is increased by 1. Here are some examples:

75 rounded to the nearest ten is 80

380 rounded to the nearest hundred is 400

22.697 rounded to the nearest hundredth is 22.70

When rounding to a certain place value, again consider the next digit after that place value. When that digit is below 5, the digit in the selected place value stays the same. The digit used to determine the rounding, and all subsequent digits, become 0. Here are some examples:

92 rounded to the nearest ten is 90

839 rounded to the nearest hundred is 800

22.64 rounded to the nearest hundredth is 22.60

Addition

Addition is the combination of two numbers so their quantities are added together cumulatively. The sign for an addition operation is the + symbol. For example, 9 + 6 = 15. The 9 and 6 combine to achieve a cumulative value, called a *sum.*

Addition holds the *commutative property*, which means that numbers in an addition equation can be switched without altering the result. The formula for the commutative property is a + b = b + a. The following examples can demonstrate how the commutative property works:

$$7 = 3 + 4 = 4 + 3 = 7$$

$$20 = 12 + 8 = 8 + 12 = 20$$

Addition also holds the associative property, which means that the grouping of numbers does not matter in an addition problem. In other words, the presence or absence of parentheses is irrelevant. The formula for the associative property is (a + b) + c = a + (b + c). Here are some examples of the associative property at work:

$$30 = (6 + 14) + 10 = 6 + (14 + 10) = 30$$

$$35 = 8 + (2 + 25) = (8 + 2) + 25 = 35$$

Subtraction

Subtraction is taking away one number from another, so their quantities are reduced. The sign designating a subtraction operation is the – symbol, and the result is called the *difference*. For example, 9 - 6 = 3. The number *6* detracts from the number *9* to reach the difference *3*.

Unlike addition, subtraction follows neither the commutative nor associative properties. The order and grouping in subtraction impact the result.

$$15 = 22 - 7 \neq 7 - 22 = -15$$

$$3 = (10 - 5) - 2 \neq 10 - (5 - 2) = 7$$

When working through subtraction problems involving larger numbers, it's necessary to regroup the numbers. The following practice problem uses regrouping:

$$\begin{array}{r} 3\ 2\ 5 \\ -\ 7\ 7 \\ \hline \end{array}$$

Here, it is clear that the ones and tens columns for 77 are greater than the ones and tens columns for 325. To subtract this number, one needs to borrow from the tens and hundreds columns. When borrowing from a column, subtracting 1 from the lender column will add 10 to the borrower column:

$$\begin{array}{rrr} 3-1 & 10+2-1 & 10+5 \\ - & 7 & 7 \\ \hline \end{array} = \begin{array}{rrr} 2 & 11 & 15 \\ - & 7 & 7 \\ \hline 2 & 4 & 8 \end{array}$$

After ensuring that each digit in the top row is greater than the digit in the corresponding bottom row, subtraction can proceed as normal, and the answer is found to be 248.

Multiplication

Multiplication involves adding together multiple copies of a number. It is indicated by an $\times$ symbol or a number immediately outside of a parenthesis. For example:

$$5(8-2)$$

The two numbers being multiplied together are called *factors*, and their result is called a *product.* For example, $9 \times 6 = 54$. This can be shown alternatively by expansion of either the 9 or the 6:

$$9 \times 6 = 9 + 9 + 9 + 9 + 9 + 9 = 54$$

$$9 \times 6 = 6 + 6 + 6 + 6 + 6 + 6 + 6 + 6 + 6 = 54$$

Like addition, multiplication holds the commutative and associative properties:

$$115 = 23 \times 5 = 5 \times 23 = 115$$

$$84 = 3 \times (7 \times 4) = (3 \times 7) \times 4 = 84$$

Multiplication also follows the distributive property, which allows the multiplication to be distributed through parentheses. The formula for distribution is $a \times (b + c) = ab + ac$. This is clear after the examples:

$$45 = 5 \times 9 = 5(3 + 6) = (5 \times 3) + (5 \times 6) = 15 + 30 = 45$$

$$20 = 4 \times 5 = 4(10 - 5) = (4 \times 10) - (4 \times 5) = 40 - 20 = 20$$

Multiplication becomes slightly more complicated when multiplying numbers with decimals. The easiest way to answer these problems is to ignore the decimals and multiply as if they were whole numbers. After multiplying the factors, a decimal gets placed in the product. The placement of the decimal is determined by taking the cumulative number of decimal places in the factors.

For example:

$$\begin{array}{r} 0.7 \\ \times\ 3 \\ \hline 2.1 \end{array} \qquad \begin{array}{r} 2.6 \\ \times\ 4.2 \\ \hline 10.92 \end{array} \qquad \begin{array}{r} 1.5 \\ \times\ 6.4 \\ \hline 9.60 \end{array}$$

Starting with the first example, the first step is to ignore the decimal and multiply the numbers as though they were whole numbers, which results in a product of 21. The next step is to count the number of digits that follow a decimal (one, in this case). Finally, the decimal place gets moved that many positions to the left, because the factors have only one decimal place. The second example works the same way, except that there are two total decimal places in the factors, so the product's decimal is moved two places over. In the third example, the decimal should be moved over two digits, but the digit zero is no longer needed, so it is erased and the final answer is 9.6.

Division

Division and multiplication are inverses of each other in the same way that addition and subtraction are opposites. The signs designating the division operation are the ÷ and / symbols. In division, the second number divides into the first.

The number before the division sign is called the *dividend* or, if expressed as a fraction, the *numerator*. For example, in $a \div b$, a is the dividend, while in $\frac{a}{b}$, a is the numerator.

The number after the division sign is called the *divisor* or, if expressed as a fraction, the *denominator*. For example, in $a \div b$, b is the divisor, while in $\frac{a}{b}$, b is the denominator.

Like subtraction, division doesn't follow the commutative property, as it matters which number comes before the division sign, and division doesn't follow the associative or distributive properties for the same reason. For example:

$$\frac{3}{2} = 9 \div 6 \neq 6 \div 9 = \frac{2}{3}$$

$$2 = 10 \div 5 = (30 \div 3) \div 5 \neq 30 \div (3 \div 5) = 30 \div \frac{3}{5} = 50$$

$$25 = 20 + 5 = (40 \div 2) + (40 \div 8) \neq 40 \div (2 + 8) = 40 \div 10 = 4$$

If a divisor doesn't divide into a dividend an integer number of times, whatever is left over is termed the *remainder*. The remainder can be further divided out into decimal form by using long division; however, this doesn't always give a quotient with a finite number of decimal places, so the remainder can also be expressed as a fraction over the original divisor.

Division with decimals is similar to multiplication with decimals in that when dividing a decimal by a whole number, one should ignore the decimal and divide as if it was a whole number.

Upon finding the answer, or quotient, the decimal point is inserted at the decimal place equal to that in the dividend.

$$15.75 \div 3 = 5.25$$

When the divisor is a decimal number, both the divisor and dividend get multiplied by 10. This process is repeated until the divisor is a whole number, then one needs to complete the division operation as described above.

$$17.5 \div 2.5 = 175 \div 25 = 7$$

Exponents

An *exponent* is an operation used as shorthand for a number multiplied or divided by itself for a defined number of times.

$$3^7 = 3 \times 3 \times 3 \times 3 \times 3 \times 3 \times 3$$

In this example, the 3 is called the *base* and the 7 is called the *exponent*. The exponent is typically expressed as a superscript number near the upper right side of the base, but can also be identified as the number following a caret symbol (^). This operation is verbally expressed as "3 to the 7th power" or "3 raised to the power of 7." Common exponents are 2 and 3. A base raised to the power of 2 is referred to as having been "squared," while a base raised to the power of 3 is referred to as having been "cubed."

Several special rules apply to exponents. First, the Zero Power Rule finds that any number raised to the zero power equals 1. For example, 100^0, 2^0, $(-3)^0$and 0^0 all equal 1 because the bases are raised to the zero power.

Second, exponents can be negative. With negative exponents, the equation is expressed as a fraction, as in the following example:

$$3^{-7} = \frac{1}{3^7} = \frac{1}{3 \text{ x } 3 \text{ x } 3 \text{ x } 3 \text{ x } 3 \text{ x } 3 \text{ x } 3}$$

Third, the Power Rule concerns exponents being raised by another exponent. When this occurs, the exponents are multiplied by each other:

$$(x^2)^3 = x^6 = (x^3)^2$$

Fourth, when multiplying two exponents with the same base, the Product Rule requires that the base remains the same and the exponents are added. For example, a^xx a^y = a^{x+y}. Since addition and multiplication are commutative, the two terms being multiplied can be in any order.

$$x^3x^5 = x^{3+5} = x^8 = x^{5+3} = x^5x^3$$

Fifth, when dividing two exponents with the same base, the Quotient Rule requires that the base remains the same, but the exponents are subtracted. So, $a^x \div a^y = a^{x-y}$. Since subtraction and division are not commutative, the two terms must remain in order.

$$x^5x^{-3} = x^{5-3} = x^5 \div x^3 = \frac{x^5}{x^3} = x^2$$

Additionally, 1 raised to any power is still equal to 1, and any number raised to the power of 1 is equal to itself. In other words, $a^1 = a$ and $14^1 = 14$.

Exponents play an important role in scientific notation to present extremely large or small numbers as follows: $a \times 10^b$. To write the number in scientific notation, the decimal is moved until there is only one digit on the left side of the decimal point, indicating that the number *a* has a value between 1 and 10. The number of times the decimal moves indicates the exponent to which 10 is raised, here represented by b. If the decimal moves to the left, then *b* is positive, but if the decimal moves to the right, then *b* is negative. The following examples demonstrate these concepts:

$$3{,}050 = 3.05 \times 10^3$$

$$-777 = -7.77 \times 10^2$$

$$0.000123 = 1.23 \times 10^{-4}$$

$$-0.0525 = -5.25 \times 10^{-2}$$

Roots

The *square root symbol* is expressed as $\sqrt{\ }$ and is commonly known as the radical. Taking the root of a number is the inverse operation of multiplying that number by itself some number of times. For example, squaring the number 7 is equal to 7 × 7, or 49. Finding the square root is the opposite of finding an exponent, as the operation seeks a number that when multiplied by itself, equals the number in the square root symbol.

For example, $\sqrt{36}$ = 6 because 6 multiplied by 6 equals 36. Note, the square root of 36 is also -6 since -6 × -6 = 36. This can be indicated using a plus/minus symbol like this: ±6. However, square roots are often just expressed as a positive number for simplicity, with it being understood that the true value can be either positive or negative.

Perfect squares are numbers with whole number square roots. The list of perfect squares begins with 0, 1, 4, 9, 16, 25, 36, 49, 64, 81, and 100.

Determining the square root of imperfect squares requires a calculator to reach an exact figure. It's possible, however, to approximate the answer by finding the two perfect squares that the number fits between. For example, the square root of 40 is between 6 and 7 since the squares of those numbers are 36 and 49, respectively.

Square roots are the most common root operation. If the radical doesn't have a number to the upper left of the symbol $\sqrt{}$, then it's a square root. Sometimes a radical includes a number in the upper left, like $\sqrt[3]{27}$, as in the other common root type—the cube root. Complicated roots, like the cube root, often require a calculator.

Parentheses

Parentheses separate different parts of an equation, and operations within them should be thought of as taking place before the outside operations take place. Practically, this means that the distinction between what is inside and outside of the parentheses decides the order of operations that the equation follows. Failing to solve operations inside the parentheses before addressing the part of the equation outside of the parentheses will lead to incorrect results.

For example, in $5 - (3 + 25)$, addition within the parentheses must be solved first. So $3 + 25 = 28$, leaving $5 - (28) = -23$. If this was solved using the incorrect order of operations, the solution might be found to be $5 - 3 + 25 = 2 + 25 = 27$, which would be wrong.

Equations often feature multiple layers of parentheses. To differentiate them, square brackets [ ] and braces { } are used in addition to parentheses. The innermost parentheses must be solved before working outward to larger brackets. For example, in $\{2 \div [5 - (3 + 1)]\}$, solving the innermost parentheses (3 + 1) leaves $\{2 \div [5 - (4)]\}$. $[5 - (4)]$ is now the next smallest, which leaves $\{2 \div [1]\}$ in the final step, and 2 as the answer.

Order of Operations

When solving equations with multiple operations, special rules apply. These rules are known as the Order of Operations. The order is as follows: Parentheses, Exponents, Multiplication and Division from left to right, and Addition and Subtraction from left to right. A popular mnemonic device to help remember the order is Please Excuse My Dear Aunt Sally (PEMDAS). Evaluating the following two problems can help with understanding the Order of Operations:

1. $4 + (3 \times 2)^2 \div 4$

   First, the operation within the parentheses must be completed, yielding: $4 + 6^2 \div 4$.

   Second, the exponent is evaluated: $4 + 36 \div 4$.

   Third, the division is conducted: $4 + 9$.

   Fourth, addition is completed, giving the answer: 13.

2. $2 \times (6+3) \div (2+1)^2$

$2 \times 9 \div (3)^2$

$2 \times 9 \div 9$

$18 \div 9$

2

Positive and Negative Numbers

Signs

Aside from 0, numbers can be either positive or negative. The sign for a positive number is the plus sign or the + symbol, while the sign for a negative number is minus sign or the – symbol. If a number has no designation, then it's assumed to be positive.

Absolute Values

Both positive and negative numbers are valued according to their distance from 0. Both +3 and -3 can be considered using the following number line:

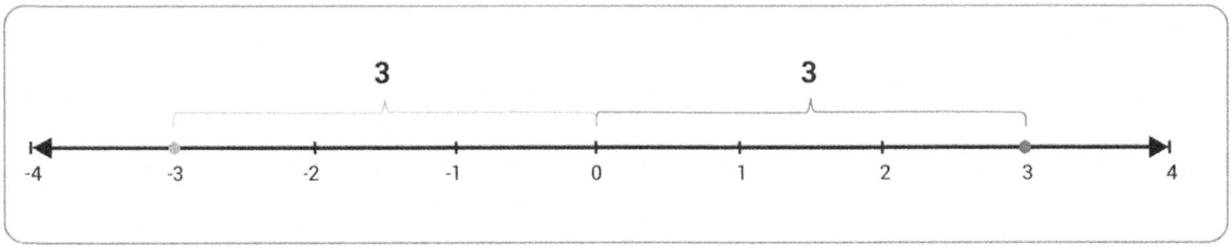

Both 3 and -3 are three spaces from 0. The distance from 0 is called its absolute value. Thus, both -3 and 3 have an absolute value of 3 since they're both three spaces away from 0.

An absolute number is written by placing | | around the number. So, $|3|$ and $|-3|$ both equal 3, as that's their common absolute value.

Implications for Addition and Subtraction

For addition, if all numbers are either positive or negative, they are simply added together. For example, 4 + 4 = 8 and -4 + -4 = -8. However, things get tricky when some of the numbers are negative and some are positive.

For example, with 6 + (-4), the first step is to take the absolute values of the numbers, which are 6 and 4. Second, the smaller value is subtracted from the larger. The equation becomes $6 - 4 = 2$. Third, the sign of the original larger number is placed on the sum. Here, 6 is the larger number, and it's positive, so the sum is 2.

Here's an example where the negative number has a larger absolute value: (-6) + 4. The first two steps are the same as the example above. However, on the third step, the negative sign must be placed on the sum, because the absolute value of (-6) is greater than 4. Thus, -6 + 4 = -2.

The absolute value of numbers implies that subtraction can be thought of as flipping the sign of the number following the subtraction sign and simply adding the two numbers. This means that subtracting a negative number will, in fact, be adding the positive absolute value of the negative number. Here are some examples:

$$-6 - 4 = -6 + -4 = -10$$

$$3 - -6 = 3 + 6 = 9$$

$$-3 - 2 = -3 + -2 = -5$$

Implications for Multiplication and Division

For multiplication and division, if both numbers are positive, then the product or quotient is always positive. If both numbers are negative, then the product or quotient is also positive. However, if the numbers have opposite signs, the product or quotient is always negative.

Simply put, the product in multiplication and quotient in division is always positive, unless the numbers have opposing signs, in which case it's negative. Here are some examples:

$$(-6) \times (-5) = 30$$

$$(-50) \div 10 = -5$$

$$8 \times |-7| = 56$$

$$(-48) \div (-6) = 8$$

If there are more than two numbers in a multiplication or division problem, then whether the product or quotient is positive or negative depends on the number of negative numbers in the problem. If there is an odd number of negatives, then the product or quotient is negative. If there is an even number of negative numbers, then the result is positive.

Here are some examples:

$$(-6) \times 5 \times (-2) \times (-4) = -240$$

$$(-6) \times 5 \times 2 \times (-4) = 240$$

Factorization

Factors are the numbers multiplied to achieve a product. Thus, every product in a multiplication equation has, at minimum, two factors. Of course, some products will have more than two factors. For the sake of most discussions, one can assume that factors are positive integers.

To find a number's factors, one should start with 1 and the number itself. The next step is to divide the number by 2, 3, 4, and so on, to see if any divisors can divide the number without a remainder. A list should be kept of those that do. This process can be stopped upon reaching either the number itself or another factor.

For example, to find the factors of 45, the first step is to start with 1 and 45. The next step is to try to divide 45 by 2, which fails. After this, 45 gets divided by 3. The answer is 15, so 3 and 15 are now factors. Dividing by 4 doesn't work, and dividing by 5 leaves 9. Lastly, dividing 45 by 6, 7, and 8 all don't work. The next integer to try is 9, but this is already known to be a factor, so the factorization is complete. The factors of 45 are 1, 3, 5, 9, 15 and 45.

Prime Factorization

Prime factorization involves an additional step after breaking a number down to its factors: breaking down the factors until they are all prime numbers. A prime number is any number that can only be divided by 1 and itself. The prime numbers between 1 and 20 are 2, 3, 5, 7, 11, 13, 17, and 19. As a simple test, numbers that are even or end in 5 are not prime.

When attempting to break 129 down into its prime factors, the factors are found first: 3 and 43. Both 3 and 43 are prime numbers, so that means the prime factorization is complete. If 43 was not a prime number, then it would also need to be factorized until all of the factors were expressed as prime numbers.

Common Factors

A common factor is a factor shared by two numbers. The following examples demonstrate how to find the common factors of 45 and 30:

- The factors of 45 are: 1, 3, 5, 9, 15, and 45.
- The factors of 30 are: 1, 2, 3, 5, 6, 10, 15, and 30.
- The common factors are 1, 3, 5, and 15.

Greatest Common Factor

The greatest common factor is the largest number among the shared, common factors. From the factors of 45 and 30, the common factors are 3, 5, and 15. Thus, 15 is the greatest common factor, as it's the largest number.

Least Common Multiple

The least common multiple is the smallest number that's a multiple of two numbers. For example, to find the least common multiple of 4 and 9, the multiples of 4 and 9 are found first. The multiples of 4 are 4, 8, 12, 16, 20, 24, 28, 32, 36, and so on. For 9, the multiples are 9, 18, 27, 36, 45, 54, etc. Thus, the least common multiple of 4 and 9 is 36, the lowest number where 4 and 9 share multiples.

If two numbers share no factors besides 1 in common, then their least common multiple will be simply their product. If two numbers have common factors, then their least common multiple will be their product divided by their greatest common factor. This can be visualized by the formula $LCM = \frac{x \times y}{GCF}$, where *x* and *y* are some integers and *LCM* and *GCF* are their least common multiple and greatest common factor, respectively.

Fractions

A fraction is an equation that represents a part of a whole, but can also be used to present ratios or division problems. An example of a fraction is $\frac{x}{y}$. In this example, x is called the *numerator,* while y is the *denominator*. The numerator represents the number of parts, and the denominator is the total number of parts. They are separated by a line or slash, known as a fraction bar.. In simple fractions, the numerator and denominator can be nearly any integer. However, the denominator of a fraction can never be zero, because dividing by zero is a function, which is undefined.

To visualize the basic idea of fractions, one can imagine that an apple pie has been baked for a holiday party, and the full pie has eight slices. After the party, there are five slices left. How could the amount of the pie that remains be expressed as a fraction? The numerator is 5 since there are five parts left, and the denominator is 8, since there were eight total slices in the whole pie. Thus, expressed as a fraction, the leftover pie totals $\frac{5}{8}$ of the original amount.

Fractions come in three different varieties: proper fractions, improper fractions, and mixed numbers. Proper fractions have a numerator less than the denominator, such as $\frac{3}{8}$, but improper fractions have a numerator greater than the denominator, such as $\frac{15}{8}$. Mixed numbers combine a whole number with a proper fraction, such as $3\frac{1}{2}$. Any mixed number can be written as an improper fraction by multiplying the integer by the denominator, adding the product to the value of the numerator, and dividing the sum by the original denominator. For example, $3\frac{1}{2} = \frac{3 \times 2 + 1}{2} = \frac{7}{2}$. Whole numbers can also be converted into fractions by placing the whole number as the numerator and making the denominator 1. For example, $3 = \frac{3}{1}$.

One of the most fundamental concepts of fractions is their ability to be manipulated by multiplication or division. This is possible since $\frac{n}{n} = 1$ for any non-zero integer. As a result, multiplying or dividing by $\frac{n}{n}$ will not alter the original fraction since any number multiplied or divided by 1 doesn't change the value of that number. Fractions of the same value are known as equivalent fractions. For example, $\frac{2}{4}, \frac{4}{8}, \frac{50}{100}$, and $\frac{75}{150}$ are equivalent, as they all equal $\frac{1}{2}$.

Although many equivalent fractions exist, they are easier to compare and interpret when reduced or simplified. The numerator and denominator of a simple fraction will have no factors in common other than 1. When reducing or simplifying fractions, the numerator and denominator are divided by the greatest common factor. A simple strategy is to divide the numerator and denominator by low numbers, like 2, 3, or 5 until arriving at a simple fraction, but the same thing could be achieved by determining the greatest common factor for both the numerator and denominator and dividing each by it. Using the first method is preferable when both the numerator and denominator are even, end in 5, or are obviously a multiple of another number. However, if no numbers seem to work, it will be necessary to factor the numerator and denominator to find the GCF. The following problems provide examples:

1. Simplify the fraction $\frac{6}{8}$:

Dividing the numerator and denominator by 2 results in $\frac{3}{4}$, which is a simple fraction.

2. Simplify the fraction $\frac{12}{36}$:

Dividing the numerator and denominator by 2 leaves $\frac{6}{18}$. This isn't a simplified fraction, as both the numerator and denominator have factors in common. Diving each by 3 results in $\frac{2}{6}$, but this can be further simplified by dividing by 2, to get $\frac{1}{3}$. This is the simplest fraction, as the numerator is 1. In cases like this, multiple division operations can be avoided by determining the greatest common factor between the numerator and denominator.

3. Simplify the fraction $\frac{18}{54}$ by dividing by the greatest common factor:

The first step is to determine the factors of the numerator and denominator. The factors of 18 are 1, 2, 3, 6, 9, and 18. The factors of 54 are 1, 2, 3, 6, 9, 18, 27, and 54. Thus, the greatest common factor is 18. Dividing $\frac{18}{54}$ by 18 leaves $\frac{1}{3}$, which is the simplest fraction. This method takes slightly more work, but it definitively arrives at the simplest fraction.

A *ratio* is a comparison between the relative sizes of two parts of a whole, separated by a colon. It's different from a fraction because, in a ratio, the second number represents the number of parts which aren't currently being referenced, while in a fraction, the second or bottom number represents the total number of parts in the whole. For example, if 3 pieces of an 8-piece pie were eaten, the number of uneaten parts expressed as a ratio to the number of eaten parts would be 5:3.

Equivalent ratios work just like equivalent fractions. For example, both 3:9 and 20:60 are equivalent ratios to 1:3 because both can be simplified to 1:3.

Operations with Fractions

Of the four basic operations that can be performed on fractions, the one that involves the least amount of work is multiplication. To multiply two fractions, the numerators are multiplied, the denominators are multiplied, and the products are placed together as a fraction. Whole numbers and mixed numbers can also be expressed as a fraction, as described above, which more easily facilitates multiplication with another fraction. The following problems provide examples:

1. $\frac{2}{5} \times \frac{3}{4} = \frac{6}{20} = \frac{3}{10}$
2. $\frac{4}{9} \times \frac{7}{11} = \frac{28}{99}$

Dividing fractions is similar to multiplication with one key difference. To divide fractions, the numerator and denominator of the second fraction are flipped, and then one proceeds as if it were a multiplication problem:

1. $\frac{7}{8} \div \frac{4}{5} = \frac{7}{8} \times \frac{5}{4} = \frac{35}{32}$
2. $\frac{5}{9} \div \frac{1}{3} = \frac{5}{9} \times \frac{3}{1} = \frac{15}{9} = \frac{5}{3}$

Addition and subtraction require more steps than multiplication and division, as these operations require the fractions to have the same denominator, also called a *common denominator*. It is always possible to find a common denominator by multiplying the denominators. However, when the denominators are large numbers, this method is unwieldy, especially if the answer must be provided in its simplest form. Thus, it's beneficial to find the least common denominator of the fractions—the least common denominator is incidentally also the least common multiple.

Once equivalent fractions have been found with common denominators, the numerators are simply added or subtracted to arrive at the answer:

1. $\frac{1}{2} + \frac{3}{4} = \frac{2}{4} + \frac{3}{4} = \frac{5}{4}$

2. $\frac{3}{12} + \frac{11}{20} = \frac{15}{60} + \frac{33}{60} = \frac{48}{60} = \frac{4}{5}$

3. $\frac{7}{9} - \frac{4}{15} = \frac{35}{45} - \frac{12}{45} = \frac{23}{45}$

4. $\frac{5}{6} - \frac{7}{18} = \frac{15}{18} - \frac{7}{18} = \frac{8}{18} = \frac{4}{9}$

Order of Rational Numbers

A common question type on the SIFT asks test takers to order rational numbers from least to greatest or greatest to least. The numbers will come in a variety of formats, including decimals, percentages, roots, fractions, and whole numbers. These questions test for knowledge of different types of numbers and the ability to determine their respective values.

Whether the question asks to order the numbers from greatest to least or least to greatest, the crux of the question is the same—convert the numbers into a common format. Generally, it's easiest to write the numbers as whole numbers and decimals so they can be placed on a number line. The following examples illustrate this strategy:

1. Order the following rational numbers from greatest to least:

$\sqrt{36}$, 0.65, 78%, $\frac{3}{4}$, 7, 90%, $\frac{5}{2}$

Of the seven numbers, the whole number (7) and decimal (0.65) are already in an accessible form, so test takers should concentrate on the other five.

First, the square root of 36 equals 6. (If the test asks for the root of a non-perfect root, determine which two whole numbers the root lies between.) Next, the percentages should be converted to decimals. A percentage means "per hundred," so this conversion requires moving the decimal point two places to the left, leaving 0.78 and 0.9. Lastly, the fractions are evaluated: $\frac{3}{4} = \frac{75}{100} = 0.75$; $\frac{5}{2} = 2\frac{1}{2} = 2.5$

Now, the only step left is to list the numbers in the requested order:

7, $\sqrt{36}$, $\frac{5}{2}$, 90%, 78%, $\frac{3}{4}$, 0.65

2. Order the following rational numbers from least to greatest:

$2.5, \sqrt{9}, -10.5, 0.853, 175\%, \sqrt{4}, \frac{4}{5}$

$\sqrt{9} = 3$

175% = 1.75

$\sqrt{4} = 2$

$\frac{4}{5} = 0.8$

From least to greatest, the answer is: $-10.5, \frac{4}{5}, 0.853, 175\%, \sqrt{4}, 2.5, \sqrt{9}$.

Percentages

Percentages can be thought of as fractions with a denominator of 100. In fact, percentage means "per hundred." Problems often require converting numbers from percentages, fractions, and decimals. The following explains how to work through those conversions.

Converting Fractions to Percentages: The fraction is converted by using an equivalent fraction with a denominator of 100. For example, $\frac{3}{4} = \frac{3}{4} \times \frac{25}{25} = \frac{75}{100} = 75\%$

Converting Percentages to Fractions: Percentages can be converted to fractions by turning the percentage into a fraction with a denominator of 100. Test takers should be wary of questions asking the converted fraction to be written in the simplest form. For example, $35\% = \frac{35}{100}$ which, although correctly written, has a numerator and denominator with a greatest common factor of 5, so it can be simplified to $\frac{7}{20}$.

Converting Percentages to Decimals: Because a percentage is based on "per hundred," decimals and percentages can be converted by multiplying or dividing by 100. Practically speaking, this always amounts to moving the decimal point two places to the right or left, depending on the conversion. To convert a percentage to a decimal, the decimal point is moved two places to the left and the % sign gets removed. To convert a decimal to a percentage, the decimal point is moved two places to the right and a "%" sign is added. Here are some examples:

65% = 0.65

0.33 = 33%

0.215 = 21.5%

99.99% = 0.9999

500% = 5.00

7.55 = 755%

Percentage Problems

Questions dealing with percentages can be difficult when they are phrased as word problems. These word problems almost always come in one of three varieties. The first type will ask to find what percentage of some number will equal another number. The second asks to determine what number is some percentage of another given number. The third will ask what number another number is a given percentage of.

One of the most important parts of correctly answering percentage word problems is to identify the numerator and the denominator. This fraction can then be converted into a percentage, as described in the previous section.

The following word problem shows how to make this conversion:

A department store carries several different types of footwear. The store is currently selling 8 athletic shoes, 7 dress shoes, and 5 sandals. What percentage of the store's footwear are sandals?

The first step is to calculate what serves as the 'whole', as this will be the denominator. How many total pieces of footwear does the store sell? The store sells 20 different types (8 athletic + 7 dress + 5 sandals).

In the next step, test takers need to determine which footwear type the question is specifically asking about. Sandals. Thus, 5 is the numerator.

Lastly, the resultant fraction must be expressed as a percentage. The first two steps indicate that $\frac{5}{20}$ of the footwear pieces are sandals. This fraction must now be converted into a percentage:

$$\frac{5}{20} \times \frac{5}{5} = \frac{25}{100} = 25\%$$

Algebra

Relations and Functions

First, it's important to understand the definition of a *relation*. Given two variables, x and y, which stand for unknown numbers, a *relation* between x and y is an object that splits all of the pairs (x, y) into those for which the relation is true and those for which it is false. For example, consider the relation of $x^2 = y^2$. This relationship is true for the pair (1, 1) and for the pair (-2, 2), but false for (2, 3). Another example of a relation is $x \leq y$. This is true whenever x is less than or equal to y.

A *function* is a special kind of relation where, for each value of x, there is only a single value of y that satisfies the relation. So, $x^2 = y^2$ is *not* a function because in this case, if x is 1, y can be either 1 or -1: the pair (1, 1) and (1, -1) both satisfy the relation. More generally, for this relation, any pair of the form $(a, \pm a)$ will satisfy the relation. On the other hand, consider the following relation: $y = x^2 + 1$. This is a function because for each value of x, there is a unique value of y that satisfies the relation. Notice, however, there are multiple values of x that give us the same value of y. This is perfectly acceptable for a function. Therefore, y is a function of x.

To determine if a relation is a function, one should check to see if every x-value has a unique corresponding y-value.

A function can be viewed as an object that has x as its input and outputs a unique *y*-value. It is sometimes convenient to express this using *function notation*, where the function itself is given a name, often *f*. To emphasize that *f* takes x as its input, the function is written as $f(x)$. In the above example, the equation could be rewritten as $f(x) = x^2 + 1$. To write the value that a function yields for some specific value of x, that value is put in place of x in the function notation. For example, $f(3)$ will denote the value that the function outputs when the input value is 3. If $f(x) = x^2 + 1$, then $f(3) = 3^2 + 1 = 10$.

A function can also be viewed as a table of pairs (x, *y*), which lists the value for *y* for each possible value of x.

The set of all possible values for x in $f(x)$ is called the *domain* of the function, and the set of all possible outputs is called the *range* of the function. Note that usually the domain is assumed to be all real numbers, except those for which the expression for $f(x)$ is not defined, unless the problem specifies otherwise. An example of how a function might not be defined is in the case of $f(x) = \frac{1}{x+1}$, which is not defined when $x = -1$ (which would require dividing by zero). Therefore, in this case, the domain would be all real numbers except $x = -1$.

If *y* is a function of x, then x is the *independent variable* and *y* is the *dependent variable*. This is because in many cases, the problem will start with some value of x and then see how *y* changes depending on this starting value.

Evaluating Functions

To evaluate functions, the given value is plugged in everywhere that the variable appears in the expression for the function. For example, find $g(-2)$ where $g(x) = 2x^2 - \frac{4}{x}$. To complete the problem, -2 is plugged in in the following way:

$$g(-2) = 2(-2)^2 - \frac{4}{-2}$$

$$2 \cdot 4 + 2$$

$$8 + 2 = 10$$

Defining Linear Equations

A function is considered *linear* if it can take the form of the equation $f(x) = ax + b$, or $y = ax + b$, for any two numbers *a* and *b*. A linear equation forms a straight line when graphed on the coordinate plane. An example of a linear function is shown below on the graph.

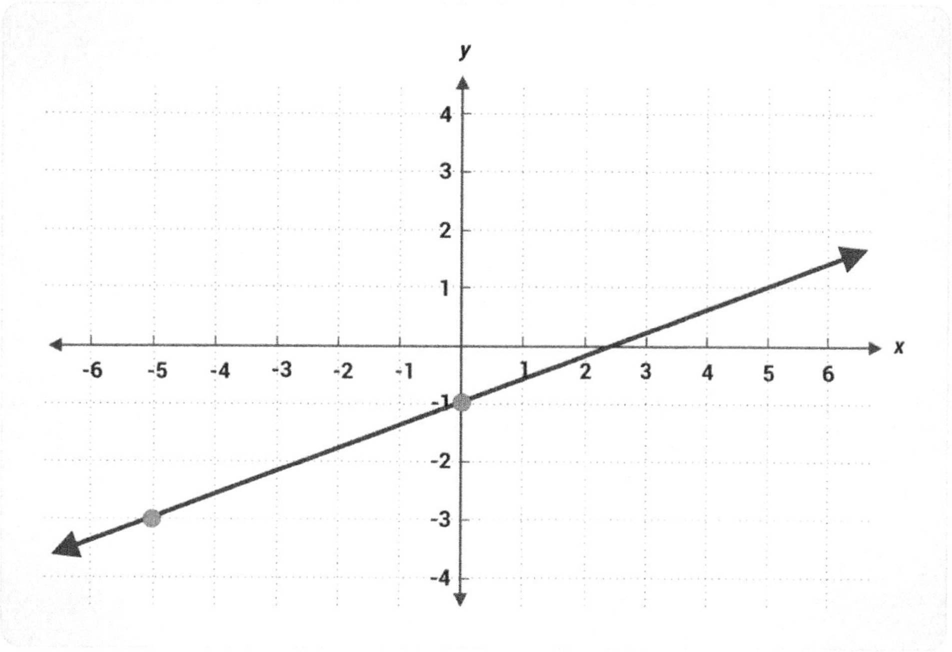

This is a graph of the function $y = \frac{2}{5}x - 1$. A table of values that satisfies this function is shown below.

x	y
-5	-3
0	-1
5	1
10	3

These points can be found on the graph using the form (x,y). For more on graphing in the coordinate plane, refer to the *Graphing* section below.

Graphing Functions and Relations

To graph relations and functions, the Cartesian plane is used. This can be visualized as a plane with a grid of squares, with one direction being the x-axis and the other direction the *y*-axis. Generally, the independent variable is placed along the horizontal axis, and the dependent variable is placed along the vertical axis. Any point on the plane can be specified by identifying the point's location along each of the two axes with a pair of numbers (x, y). Specific values for these pairs can be given names such as $C = (-1, 3)$. Negative values mean to move left or down; positive values mean to move right or up. The point where the axes cross one another is called the *origin*. The origin has coordinates $(0, 0)$ and is usually called O when given a specific label. An illustration of the Cartesian plane, along with graphs of $(2,1)$ and $(-1,-1)$, are below.

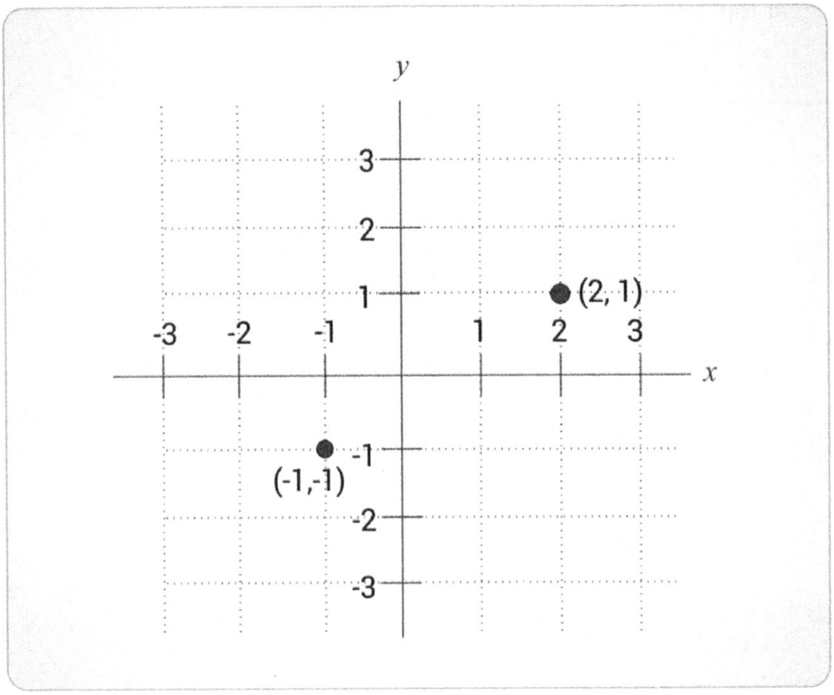

Relations also can be graphed by marking each point whose coordinates satisfy the relation. If the relation is a function, then there is only one value of y for any given value of x. This leads to the **vertical line test**: if a relation is graphed, then the relation is a function if any possible vertical line drawn anywhere along the graph would only touch the graph of the relation in no more than one place. Conversely, when graphing a function, then any possible vertical line drawn will not touch the graph of the function at any point or will touch the function at just one point. This test is made from the definition of a function, where each x-value must be mapped to one and only one y-value.

Forms of Linear Equations

When graphing a linear function, the ratio of the change of the y-coordinate to the change in the x-coordinate is constant between any two points on the resulting line, no matter which two points are chosen. In other words, in a pair of points on a line, (x_1, y_1) and (x_2, y_2), with $x_1 \neq x_2$ so that the two points are distinct, then the ratio $\frac{y_2-y_1}{x_2-x_1}$ will be the same, regardless of which particular pair of points are chosen. This ratio, $\frac{y_2-y_1}{x_2-x_1}$, is called the *slope* of the line and is frequently denoted with the letter m. If slope m is positive, then the line goes upward when moving to the right, while if slope m is negative, then the line goes downward when moving to the right. If the slope is 0, then the line is called *horizontal*, and the y-coordinate is constant along the entire line. In lines where the x-coordinate is constant along the entire line, y is not actually a function of x. For such lines, the slope is not defined. These lines are called *vertical* lines.

Linear functions may take forms other than $y = ax + b$. The most common forms of linear equations are explained below:

1. Standard Form: $Ax + By = C$, in which the slope is given by $m = \frac{-A}{B}$, and the y-intercept is given by $\frac{C}{B}$.
2. Slope-Intercept Form: $y = mx + b$, where the slope is m and the y-intercept is b.
3. Point-Slope Form: $y - y_1 = m(x - x_1)$, where the slope is m and (x_1, y_1) is any point on the chosen line.
4. Two-Point Form: $\frac{y-y_1}{x-x_1} = \frac{y_2-y_1}{x_2-x_1}$, where (x_1, y_1) and (x_2, y_2) are any two distinct points on the chosen line. Note that the slope is given by $m = \frac{y_2-y_1}{x_2-x_1}$.
5. Intercept Form: $\frac{x}{x_1} + \frac{y}{y_1} = 1$, in which x_1 is the x-intercept and y_1 is the y-intercept.

These five ways to write linear equations are all useful in different circumstances. Depending on the given information, it may be easier to write one of the forms over another.

If $y = mx$, y is directly proportional to x. In this case, changing x by a factor changes y by that same factor. If $y = \frac{m}{x}$, y is inversely proportional to x. For example, if x is increased by a factor of 3, then y will be decreased by the same factor, 3.

Solving Linear Equations

Sometimes, rather than a situation where there's an equation such as $y = ax + b$ and the goal is to find y for some value of x, the result is given and finding x is requested.

The key to solving any equation is to remember that from one true equation, another true equation can be found by adding, subtracting, multiplying, or dividing both sides by the same quantity. In this case, it's necessary to manipulate the equation so that one side only contains x. Then the other side will show what x is equal to.

For example, in solving $3x - 5 = 2$, adding 5 to each side results in $3x = 7$. Next, dividing both sides by 3 results in $x = \frac{7}{3}$. To ensure the value of x is correct, the number can be substituted into the original equation and solved to see if it makes a true statement. For example, $3(\frac{7}{3}) - 5 = 2$ can be simplified by cancelling out the two 3s. This yields $7 - 5 = 2$, which is a true statement.

Sometimes an equation may have more than one x term. For example, consider the following equation:

$$3x + 2 = x - 4$$

Moving all of the x terms to one side by subtracting x from both sides results in:

$$2x + 2 = -4$$

Next, 2 is subtracted from both sides so that there is no constant term on the left side. This yields $2x = -6$. Finally, both sides are divided by 2, which leaves $x = -3$.

Solving Linear Inequalities

Solving linear inequalities is very similar to solving equations, except for one rule: when multiplying or dividing an inequality by a negative number, the inequality symbol changes direction. Given the following inequality, solve for x: $-2x + 5 < 13$. The first step in solving this equation is to subtract 5 from both sides. This leaves the inequality: $-2x < 8$. The last step is to divide both sides by -2. By using the rule, the answer to the inequality is $x > -4$.

Since solutions to inequalities include more than one value, number lines are used many times to model the answer. For the previous example, the answer is modelled on the number line below. It shows that any number greater than -4, not including -4, satisfies the inequality.

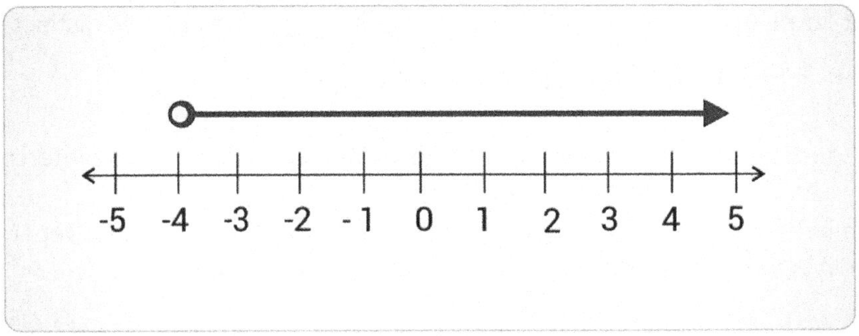

Linear Systems of Equations

A problem sometimes involves multiple variables and multiple equations. These are called *systems of equations*. In this case, one should try to manipulate them until an expression is found that provides the value of one of the variables. There are a couple of different approaches to this, and in some cases, some of them can be used together. The three basic rules to keep in mind are the following:

1. A set of equations can be manipulated by performing the same operation to both equations, just as is done when working with just one equation.

2. If one of the equations can be changed so that it expresses one variable in terms of the others, then that expression can be substituted into the other equations and the variable can be eliminated. This means the other equations will have one less variable in them. This is called the *method of substitution*.

3. If two equations of the form $a = b, c = d$ are included, then a new equation can be formed by adding the left sides and adding the right sides, $a + c = b + d$, or $a - c = b - d$. This enables the elimination of one of the variables from an equation. This is called the *method of elimination.*

The simplest case is the case of a *linear* system of equations. Although the equations may be written in more complicated forms, linear systems of equations with two variables can always be written in the form $ax + by = c, dx + ey = f$. The two basic approaches to solving these systems are substitution and elimination.

Consider the system $3x - y = 2$ and $2x + 2y = 3$. This can be solved in two ways.

1. By substitution: start by solving the first equation for y. First, subtract $3x$ from both sides to obtain $-y = 2 - 3x$. Next, divide both sides by -1, to obtain $y = 3x - 2$. Then substitute this value for y into the second equation. This yields:

$$2x + 2(3x - 2) = 3$$

This can be simplified to $2x + 6x - 4 = 3$, or $8x = 7$, which means $x = \frac{7}{8}$. Plugging in this value for x into $y = 3x - 2$, yields:

$$y = 3\left(\frac{7}{8}\right) - 2 = \frac{21}{8} - \frac{16}{8} = \frac{5}{8}$$

So, this results in $x = \frac{7}{8}, y = \frac{5}{8}$.

2. By elimination: first, multiply the first equation by 2. This results in $-2y$, which could cancel out the +$2y$ in the second equation. Multiplying both sides of the first equation by 2 gives results in $2(3x - y) = 2(2)$, or $6x - 2y = 4$. Adding the left sides and the right sides of the two equations and setting the results equal to one another results in:

$$(6x + 2x) + (-2y + 2y) = 4 + 3$$

This simplifies to $8x = 7$, so again $x = \frac{7}{8}$. Plug this back into either of the original equations and the result is $3\left(\frac{7}{8}\right) - y = 2$ or $y = 3\left(\frac{7}{8}\right) - 2 = \frac{21}{8} - \frac{16}{8} = \frac{5}{8}$. This again yields $x = \frac{7}{8}, y = \frac{5}{8}$.

As this shows, both methods will give the same answer. However, one method is sometimes preferred over another simply because of the amount of work required. To check the answer, the values can be substituted into the given system to make sure they form two true statements.

Algebraic Expressions and Equations

Algebraic expressions look similar to equations, but they do not include the equals sign. Algebraic expressions are comprised of numbers, variables, and mathematical operations. Some examples of algebraic expressions are $8x + 7y - 12z$, $3a^2$, and $5x^3 - 4y^4$.

Algebraic expressions and equations can be used to represent real-life situations and model the behavior of different variables. For example, $2x + 5$ could represent the cost to play games at an arcade. In this case, 5 represents the price of admission to the arcade and 2 represents the cost of each game played. To calculate the total cost, the number of games played is used for x. Then this value is multiplied by 2, and lastly, 5 is added to that product.

Word Problems and Applications

In word problems, multiple quantities are often provided with a request to find some kind of relation between them. This often will mean that one variable (the dependent variable whose value needs to be found) can be written as a function of another variable (the independent variable whose value can be figured from the given information). The usual procedure for solving these problems is to start by assigning each quantity in the problem a variable, and then figuring the relationship between these variables.

For example, suppose a car gets 25 miles per gallon. How far will the car travel if it uses 2.4 gallons of fuel? In this case, *y* would be the distance the car has traveled in miles, and *x* would be the amount of fuel burned in gallons (2.4). Then the relationship between these variables can be written as an algebraic equation, $y = 25x$. In this case, the equation is $y = 25 \cdot 2.4 = 60$, so the car has traveled 60 miles.

Some word problems require more than just one simple equation to be written and solved. Consider the following situations and the linear equations used to model them.

Suppose Margaret is 2 miles to the east of John at noon. Margaret walks to the east at 3 miles per hour. How far apart will they be at 3 p.m.? To solve this, *x* would represent the time in hours past noon, and *y* would represent the distance between Margaret and John. Now, noon corresponds to the equation where *x* is 0, so the *y*-intercept is going to be 2. It's also known that the slope will be the rate at which the distance is changing, which is 3 miles per hour. This means that the slope will be 3 (test takers should be careful at this point: if units other than miles and hours were used for the *x* and *y* variables, a conversion of the given information to the appropriate units would be required first). The simplest way to write an equation given the *y*-intercept, and the slope is the Slope-Intercept form, is $y = mx + b$, where *m* is the slope, and b is the *y*-intercept. So, $m = 3$ and $b = 2$. Therefore, the equation will be $y = 3x + 2$. The word problem asks how far to the east Margaret will be from John at 3 p.m., which means when *x* is 3. So, $x = 3$ is substituted into this equation to obtain:

$$y = 3 \cdot 3 + 2 = 9 + 2 = 11$$

Therefore, she will be 11 miles to the east of him at 3 p.m.

For another example, suppose that a box with 4 cans in it weighs 6 lbs., while a box with 8 cans in it weighs 12 lbs. Find out how much a single can weighs. To do this, let *x* denote the number of cans in the box, and let *y* denote the weight of the box with the cans in pounds. This line touches two pairs: $(4, 6)$ and (8, 12).

A formula for this relation could be written using the two-point for: $x_1 = 4, y_1 = 6, x_2 = 8, y_2 = 12$. This would yield $\frac{y-6}{x-4} = \frac{12-6}{8-4}$, or $\frac{y-6}{x-4} = \frac{6}{4} = \frac{3}{2}$.

However, only the slope is needed to solve this problem, since the slope will be the weight of a single can. From the computation, the slope is $\frac{3}{2}$. Therefore, each can weighs $\frac{3}{2}$ lb.

Polynomials

An expression of the form ax^n, where *n* is a non-negative integer, is called a *monomial* because it contains one term. A sum of monomials is called a *polynomial.* For example, $-4x^3 + x$ is a polynomial, while $5x^7$ is a monomial. A function equal to a polynomial is called a *polynomial function*.

The monomials in a polynomial are also called the *terms* of the polynomial.

The constants that precede the variables are called *coefficients*.

The highest value of the exponent of *x* in a polynomial is called the *degree* of the polynomial. So, $-4x^3 + x$ has a degree of 3, while $-2x^5 + x^3 + 4x + 1$ has a degree of 5. When multiplying polynomials, the degree of the result will be the sum of the degrees of the two polynomials being multiplied.

To add polynomials, add the coefficients of like powers of x. For example:

$$(-2x^5 + x^3 + 4x + 1) + (-4x^3 + x)$$

$$-2x^5 + (1 - 4)x^3 + (4 + 1)x + 1$$

$$-2x^5 - 3x^3 + 5x + 1$$

Likewise, subtraction of polynomials is performed by subtracting coefficients of like powers of x. So:

$$(-2x^5 + x^3 + 4x + 1) - (-4x^3 + x)$$

$$-2x^5 + (1 + 4)x^3 + (4 - 1)x + 1$$

$$-2x^5 + 5x^3 + 3x + 1$$

To multiply two polynomials, each term of the first polynomial is multiplied by each term of the second polynomial and then the results are added.

For example:

$$(4x^2 + x)(-x^3 + x)$$

$$4x^2(-x^3) + 4x^2(x) + x(-x^3) + x(x)$$

$$-4x^5 + 4x^3 - x^4 + x^2$$

In the case where each polynomial has two terms, like in this example, some students find it helpful to remember this as multiplying the First terms, then the Outer terms, then the Inner terms, and finally the Last terms, with the mnemonic FOIL. For longer polynomials, the multiplication process is the same, but there will be, of course, more terms, and there is no common mnemonic to remember each combination.

The process of *factoring* a polynomial means to write the polynomial as a product of other (generally simpler) polynomials. Here is an example:

$$x^2 - 4x + 3 = (x - 1)(x - 3)$$

If a certain monomial divides every term of the polynomial, it gets factored out of each term.

For example:

$$4x^3 + 16x^2 = 4x^2(x + 4).$$

$$x^2 + 2xy + y^2 = (x + y)^2 \text{ or } x^2 - 2xy + y^2 = (x - y)^2$$

$$x^2 - y^2 = (x + y)(x - y)$$

$$x^3 + y^3 = (x + y)(x^2 - xy + y^2)$$

$$x^3 - y^3 = (x - y)(x^2 + xy + y^2)$$

$$x^3 + 3x^2y + 3xy^2 + y^3 = (x + y)^3 \text{ and } x^3 - 3x^2y + 3xy^2 - y^3 = (x - y)^3$$

Sometimes, it can be necessary to rewrite the polynomial in some clever way before applying the above rules.

Consider the problem of factoring $x^4 - 1$. This does not immediately look like any of the cases for which there are rules.

However, it's possible to think of this polynomial as: $x^4 - 1 = (x^2)^2 - (1^2)^2$, which can be simplified to:

$$(x^2)^2 - (1^2)^2 = (x^2 + 1^2)(x^2 - 1^2) = (x^2 + 1)(x^2 - 1)$$

Quadratic Functions

A polynomial of degree 2 is called *quadratic*. Every quadratic function can be written in the form $ax^2 + bx + c$. The graph of a quadratic function, $y = ax^2 + bx + c$, is called a *parabola*. Parabolas are vaguely U-shaped.

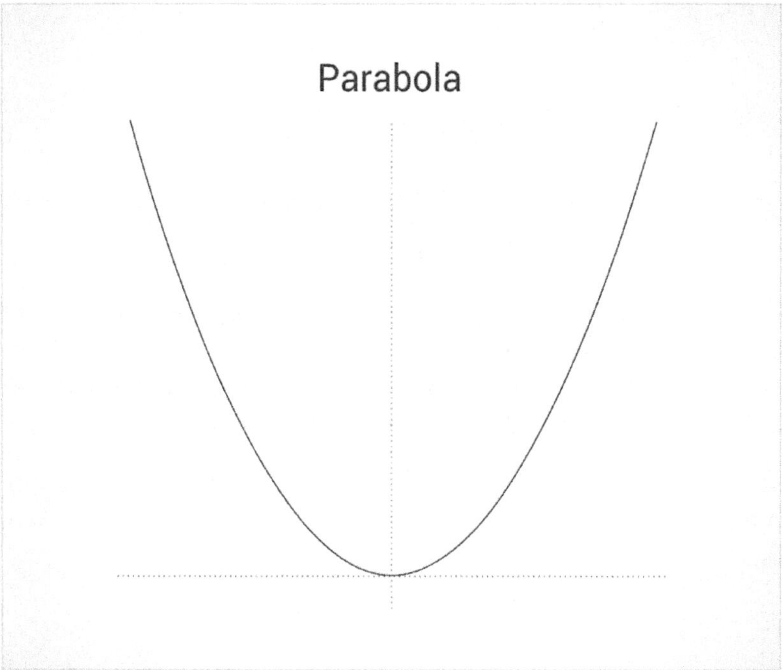

Whether the parabola opens upward or downward depends on the sign of *a*. If *a* is positive, then the parabola will open upward. If *a* is negative, then the parabola will open downward. The value of *a* will also affect how wide the parabola is. If the absolute value of *a* is large, then the parabola will be fairly skinny. If the absolute value of *a* is small, then the parabola will be quite wide.

Changes to the value of *b* affect the parabola in different ways, depending on the sign of *a*. For positive values of *a*, increasing *b* will move the parabola to the left, and decreasing *b* will move the parabola to the right. On the other hand, if *a* is negative, the effects will be the opposite: increasing *b* will move the parabola to the right, while decreasing *b* will move the parabola to the left.

Changes to the value of *c* move the parabola vertically. The larger that *c* is, the higher the parabola gets. This does not depend on the value of *a*.

The quantity $D = b^2 - 4ac$ is called the *discriminant* of the parabola. When the discriminant is positive, then the parabola has two real zeros, or x-intercepts. However, if the discriminant is negative, then there are no real zeros, and the parabola will not cross the x-axis. The highest or lowest point of the parabola is called the *vertex*. If the discriminant is zero, then the parabola's highest or lowest point is on the x-axis, and it will have a single real zero. The x-coordinate of the vertex can be found using the equation $x = -\frac{b}{2a}$. This x-value can be plugged into the equation to find the y-coordinate.

A quadratic equation is often used to model the path of an object thrown into the air. The x-value can represent the time in the air, while the y-value can represent the height of the object. In this case, the maximum height of the object would be the y-value found when the x-value is $-\frac{b}{2a}$.

Solving Quadratic Equations

A *quadratic equation* is an equation in the form $ax^2 + bx + c = 0$. There are several methods to solve such equations. The easiest method will depend on the particular quadratic equation in question.

It is sometimes possible to solve quadratic equations by manually *factoring* them. This means rewriting them in the form $(x + A)(x + B) = 0$. If this is done, then they can be solved by remembering that when $ab = 0$, either *a* or *b* must be equal to zero. Therefore, to have $(x + A)(x + B) = 0$, $(x + A) = 0$ or $(x + B) = 0$ is needed. These equations have the solutions $x = -A$ and $x = -B$, respectively.

In order to factor a quadratic equation, note that:

$$(x + A)(x + B) = x^2 + (A + B)x + AB$$

So, if an equation is in the form $x^2 + bx + c$, two numbers, *A* and *B*, need to be found that will add up to *b*, and multiply together to give *c*.

As an example, consider solving the equation:

$$-3x^2 + 6x + 9 = 0$$

The first step is to divide both sides by -3, which yields:

$$x^2 - 2x - 3 = 0$$

Because $1 - 3 = -2$ and $(1)(-3) = -3$, the equation can be factored into:

$$(x + 1)(x - 3) = 0$$

Now, one can solve $(x + 1) = 0$ and $(x - 3) = 0$ to get the solutions $x = -1$ and $x = 3$.

When trying to factor, it is useful to remember that:

$$x^2 + 2xy + y^2 = (x + y)^2,$$

$$x^2 - 2xy + y^2 = (x - y)^2$$

$$x^2 - y^2 = (x + y)(x - y)$$

However, factoring by hand is often hard to do. If there are no obvious ways to factor the quadratic equation, solutions can still be found by using the *quadratic formula*.

The quadratic formula is:

$$x = \frac{-b \pm \sqrt{b^2 - 4ac}}{2a}$$

This method will always work, although it sometimes can take longer than factoring by hand, which can be quick if the factors are easy to guess. Using the standard form $ax^2 + bx + c = 0$, the values of *a*, *b*, and *c* from the equation can be plugged into the formula to solve for x. There will either be two answers, one answer, or no real answer. No real answer comes when the value of the *discriminant*—the number under the square root—is a negative number. Since there are no real numbers that, when squared, result in a negative, the answer will be no real roots.

Here is an example of solving a quadratic equation using the quadratic formula. Suppose the equation to solve is:

$$-2x^2 + 3x + 1 = 0$$

There is no obvious way to factor this, so the quadratic formula is used, with $a = -2, b = 3, c = 1$. After substituting these values into the quadratic formula, it yields:

$$x = \frac{-3 \pm \sqrt{3^2 - 4(-2)(1)}}{2(-2)}$$

This can be simplified to obtain:

$$\frac{3 \pm \sqrt{9+8}}{4} \text{ or } \frac{3 \pm \sqrt{9+8}}{4}.$$

Challenges can be encountered when asked to find a quadratic equation with specific roots. Given roots *A* and *B*, a quadratic function can be constructed with those roots by taking:

$$(x - A)(x - B)$$

So, constructing a quadratic equation with roots $x = -2, 3$, would result in:

$$(x + 2)(x - 3) = x^2 - x - 6$$

Multiplying this by a constant also could be done without changing the roots.

Exponents and Roots

Although previously discussed, additional review regarding exponents and roots is provided.

An *exponent* is written as a^b. In this expression, *a* is called the *base* and *b* is called the *exponent*. It is properly stated that *a* is raised to the *n*-th power. Therefore, in the expression 2^3, the exponent is 3, while the base is 2. Such an expression is called an *exponential expression*. Note that when the exponent is 2, it is called *squaring* the base, and when it is 3, it is called *cubing* the base.

When the exponent is a positive integer, it indicates that the base is multiplied by itself the number of times written in the exponent. So, in the expression 2^3, 2 is multiplied by itself with 3 copies of 2: $2^3 = 2 \cdot 2 \cdot 2 = 8$. One thing to notice is that, for positive integers *n* and *m*, $a^n a^m = a^{n+m}$ is a rule. In order to make this rule be true for an integer of 0, $a^0 = 1$, so that $a^n a^0 = a^{n+0} = a^n$. And, in order to make this rule be true for negative exponents, $a^{-n} = \frac{1}{a^n}$.

Another rule for simplifying expressions with exponents is shown by the following equation: $(a^m)^n = a^{mn}$. This is true for fractional exponents as well. So, for a positive integer, $a^{\frac{1}{n}}$ is defined to be the number that, when raised to the *n*-th power, provides *a*. In other words, $(a^{\frac{1}{n}})^n = a$ is the desired equation. It should be noted that $a^{\frac{1}{n}}$ is the *n*-th root of *a*. This also can be written as $a^{\frac{1}{n}} = \sqrt[n]{a}$. The symbol on the right-hand side of this equation is called a *radical*. If the root is left out, one should assume that the 2nd root should be taken, also called the *square* root: $a^{\frac{1}{2}} = \sqrt[2]{a} = \sqrt{a}$. Additionally, $\sqrt[3]{a}$ is also called the *cube* root.

Note that when multiple roots exist, $a^{\frac{1}{n}}$ is defined to be the *positive* root. So, $4^{\frac{1}{2}} = 2$. Also note that negative numbers do not have even roots in the real numbers.

This also enables finding exponents for any rational number:

$$a^{\frac{m}{n}} = (a^{\frac{1}{n}})^m = (a^m)^{\frac{1}{n}}$$

In fact, the exponent can be any real number. In general, the following rules for exponents should be used for any numbers $a, b, m,$ and n.

- $a^1 = a.$
- $1^a = 1.$
- $a^0 = 1.$
- $a^m a^n = a^{m+n}.$
- $\frac{a^m}{a^n} = a^{m-n}$
- $(a^m)^n = a^{m \times n}.$
- $(ab)^m = a^m b^m.$
- $(\frac{a}{b})^m = \frac{a^m}{b^m}.$

Consider the problem of simplifying the expression: $(3x^2y)^3(2xy^4)$.

Start by simplifying the left term using the sixth rule listed. Applying this rule yields the following expression: $27x^6y^3(2xy^4)$.

The exponents can now be combined with base *x* and the exponents with base *y*.

The coefficients are multiplied to yield $54x^7y^7$.

Solving Equations with Exponents and Roots

Here are some of the most important properties of exponents and roots: if *n* is an integer, and if $a^n = b^n$, then $a = b$ if *n* is odd; but $a = \pm b$ if *n* is even. Similarly, if the roots of two things are equal, $\sqrt[n]{a} = \sqrt[n]{b}$, then $a = b$. This means that when starting with a true equation, both sides of that equation can be raised to a given power to obtain another true equation.

It is important to note that when an even-powered root is taken on both sides of the equation, a $\pm$ in the result. For example, given the equation $x^2 = 16$, the square root of both sides is taken to solve for x. This results in the answer $x = \pm 4$ because $(-4)^2 = 16$ and $(4)^2 = 16$.

Another property is that if $a^n = a^m$, then $n = m$. This is true for any real numbers *n* and *m*.

For solving the equation $\sqrt{x+2} - 1 = 3$, the -1 over is first moved to the right-hand side by adding 1 to both sides, which yields $\sqrt{x+2} = 4$.

Next, both sides are squared, remembering that by squaring both sides, the signs are irrelevant.

This yields: $x + 2 = 16$, which simplifies to: $x = 14$.

Now consider the problem $(x+1)^4 = 16$.

To solve this, the 4th root of both sides is taken, which means an ambiguity in the sign will be introduced because it is an even root:

$$\sqrt[4]{(x+1)^4} = \pm\sqrt[4]{16}$$

The right-hand side is 2, since $2^4 = 16$.

Therefore, $x + 1 = \pm 2$ or $x = -1 \pm 2.\backslash$

Thus, the two possible solutions are $x = -3$ and $x = 1$.

Remember that when solving equations, the answer can be checked by plugging the solution back into the problem to make a true statement.

Rational Expressions

A *rational expression* is a fraction where the numerator and denominator are both polynomials. Some examples of rational expressions include the following: $\frac{4x^3y^5}{3z^4}$, $\frac{4x^3+3x}{x^2}$, and $\frac{x^2+7x+10}{x+2}$. Since these refer to expressions and not equations, they can be simplified but not solved. Using the rules in the previous Exponents and Roots sections, some rational expressions with monomials can be simplified. Other rational expressions such as the last example, $\frac{x^2+7x+10}{x+2}$, require more steps to be simplified. First, the polynomial on top can be factored from $x^2 + 7x + 10$ into $(x+5)(x+2)$. Then the common factors can be canceled and the expression can be simplified to $(x+5)$.

The following problem is an example of using rational expressions:

Reggie wants to lay sod in his rectangular backyard. The length of the yard is given by the expression $4x + 2$ and the width is unknown. The area of the yard is $20x + 10$. Reggie needs to find the width of the yard. Knowing that the area of a rectangle is length multiplied by width, an expression can be written to find the width: $\frac{20x+10}{4x+2}$, area divided by length. Simplifying this expression by factoring out 10 on the top and 2 on the bottom leads to this expression: $\frac{10(2x+1)}{2(2x+1)}$. Cancelling out the $2x + 1$ results in $\frac{10}{2} = 5$. The width of the yard is found to be 5 by simplifying the rational expression.

Rational Equations

A *rational equation* can be as simple as an equation with a ratio of polynomials, $\frac{p(x)}{q(x)}$, set equal to a value, where $p(x)$ and $q(x)$ are both polynomials. A rational equation has an equal sign, which is different from expressions. This leads to solutions, or numbers that make the equation true.

It is possible to solve rational equations by trying to get all of the x terms out of the denominator and then isolating them on one side of the equation. For example, to solve the equation $\frac{3x+2}{2x+3} = 4$, both sides get multiplied by $(2x + 3)$. This will cancel on the left side to yield $3x + 2 = 4(2x + 3)$, then $3x + 2 = 8x + 12$. Now, subtract $8x$ from both sides, which yields $-5x + 2 = 12$. Subtracting 2 from both sides results in $-5x = 10$. Finally, both sides get divided by -5 to obtain $x = -2$.

Sometimes, when solving rational equations, it can be easier to try to simplify the rational expression by factoring the numerator and denominator first, then cancelling out common factors. For example, to solve $\frac{2x^2-8x+6}{x^2-3x+2} = 1$, the first step is to factor:

$$2x^2 - 8x + 6 = 2(x^2 - 4x + 3)$$

$$2(x-1)(x-3)$$

Then, factor $x^2 - 3x + 2$ into $(x - 1)(x - 2)$. This turns the original equation into:

$$\frac{2(x-1)(x-3)}{(x-1)(x-2)} = 1$$

The common factor of $(x - 1)$ can be canceled, leaving $\frac{2(x-3)}{x-1} = 1$. Now the same method used in the previous example can be followed. Multiplying both sides by $x - 1$ and performing the multiplication on the left yields $2x - 6 = x - 1$, which can be simplified to $x = 5$.

Rational Functions

A *rational function* is similar to an equation, but it includes two variables. In general, a rational function is in the form: $f(x) = \frac{p(x)}{q(x)}$, where $p(x)$ and $q(x)$ are polynomials. Refer to the *Functions* section for a more detailed definition of functions. Rational functions are defined everywhere except where the denominator is equal to zero. When the denominator is equal to zero, this indicates either a hole in the graph or an asymptote. An example of a function with an asymptote is shown below.

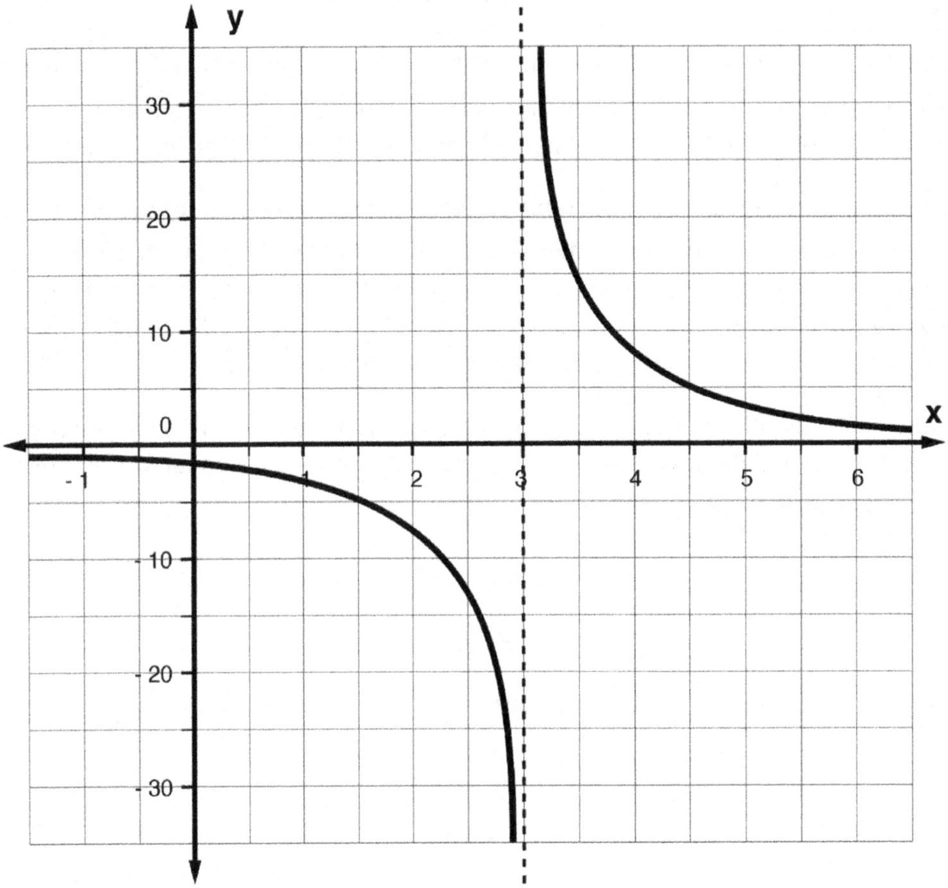

Geometry

Plane Geometry

Locations on the plane that have no width or breadth are called *points*. These points usually will be denoted with capital letters such as *P*.

Any pair of points *A*, *B* on the plane will determine a unique straight line between them. This line is denoted *AB*. Sometimes to emphasize a line is being considered, it will be written as $\overleftrightarrow{AB}$.

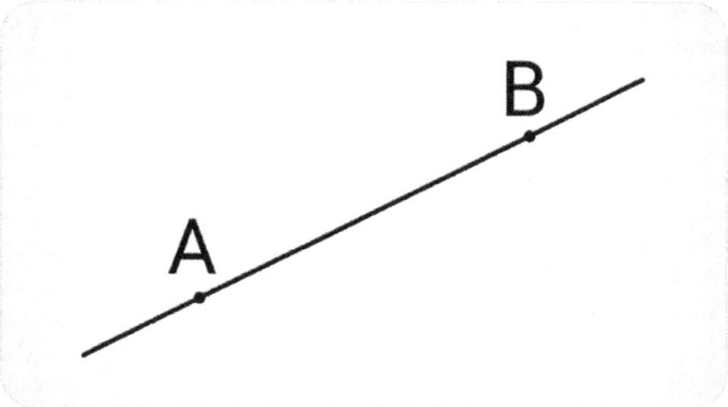

If the Cartesian coordinates for *A* and *B* are known, then the distance $d(A, B)$ along the line between them can be measured using the *Pythagorean formula*, which states that if $A = (x_1, y_1)$ and $B = (x_2, y_2)$, then the distance between them is:

$$d(A, B) = \sqrt{(x_2 - x_1)^2 + (y_2 - y_1)^2}$$

The part of a line that lies between *A* and *B* is called a *line segment*. It has two endpoints, one at *A* and one at *B*. *Rays* also can be formed. Given points *A* and *B*, a *ray* is the portion of a line that starts at one of these points, passes through the other, and keeps on going. Therefore, a ray has a single endpoint, but the other end goes off to infinity.

Given a pair of points *A* and *B*, a circle centered at *A* and passing through *B* can be formed. This is the set of points whose distance from *A* is exactly $d(A, B)$. The radius of this circle will be $d(A, B)$.

The *circumference* of a circle is the distance traveled by following the edge of the circle for one complete revolution, and the length of the circumference is given by $2\pi r$, where *r* is the radius of the circle. The formula for circumference is $C = 2\pi r$.

When two lines cross, they form an *angle*. The point where the lines cross is called the *vertex* of the angle. The angle can be named by either just using the vertex, $\angle A$, or else by listing three points $\angle BAC$, as shown in the diagram below.

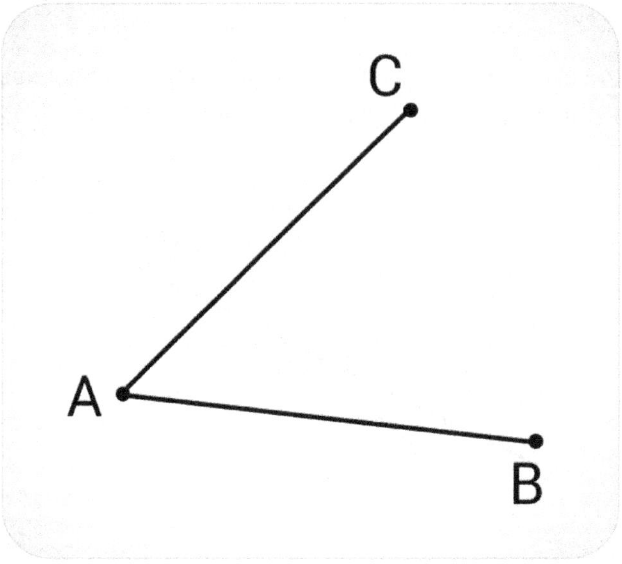

The measurement of an angle can be given in degrees or in radians. In degrees, a full circle is 360 degrees, written 360°. In radians, a full circle is 2π radians.

Given two points on the circumference of a circle, the path along the circle between those points is called an *arc* of the circle. For example, the arc between *B* and *C* is denoted by a thinner line:

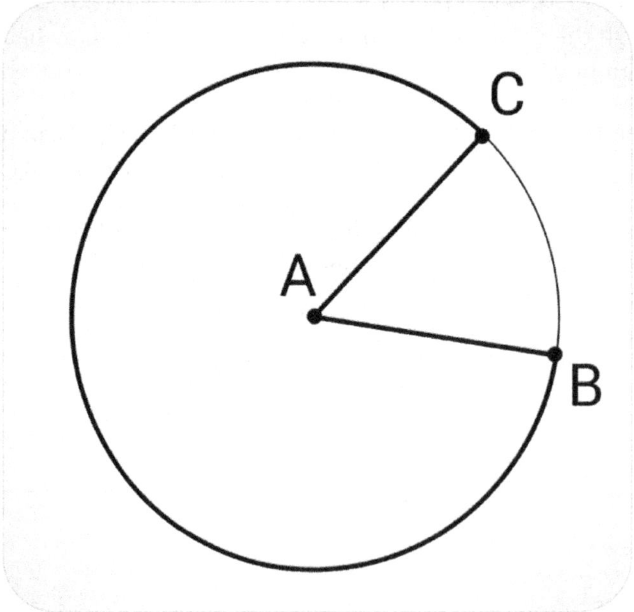

The length of the path along an arc is called the *arc length*. If the circle has radius *r*, then the arc length is given by multiplying the measure of the angle in radians by the radius of the circle.

Two lines are said to be *parallel* if they never intersect. If the lines are *AB* and *CD*, then this is written as $AB \parallel CD$.

If two lines cross to form four quarter-circles, that is, 90° angles, the two lines are *perpendicular*. If the point at which they cross is *B*, and the two lines are *AB* and *BC*, then this is written as $AB \perp BC$.

A *polygon* is a closed figure (meaning it divides the plane into an inside and an outside) consisting of a collection of line segments between points. These points are called the *vertices* of the polygon. These line segments must not overlap one another. Note that the number of sides is equal to the number of angles, or vertices of the polygon. The angles between line segments meeting one another in the polygon are called *interior angles*.

A *regular polygon* is a polygon whose edges are all the same length and whose interior angles are all of equal measure.

A *triangle* is a polygon with three sides. A *quadrilateral* is a polygon with four sides.

A *right triangle* is a triangle that has one 90° angle.

The sum of the interior angles of any triangle must add up to 180°.

An *isosceles triangle* is a triangle in which two of the sides are the same length. In this case, it will always have two congruent interior angles. If a triangle has two congruent interior angles, it will always be isosceles.

An *equilateral triangle* is a triangle whose sides are all the same length and whose angles are all equivalent to one another, equal to 60°. Equilateral triangles are examples of regular polygons. Note that equilateral triangles are also isosceles.

A *rectangle* is a quadrilateral whose interior angles are all 90°. A rectangle has two sets of sides that are equal to one another.

A *square* is a rectangle whose width and height are equal. Therefore, squares are regular polygons.

A *parallelogram* is a quadrilateral in which the opposite sides are parallel and equivalent to each other.

Transformations of a Plane

Given a figure drawn on a plane, many changes can be made to that figure, including *rotation*, *translation*, and *reflection*. Rotations turn the figure about a point, translations slide the figure, and reflections flip the figure over a specified line. When performing these transformations, the original figure is called the *pre-image*, and the figure after transformation is called the *image*.

More specifically, *translation* means that all points in the figure are moved in the same direction by the same distance. In other words, the figure is slid in some fixed direction. Of course, while the entire figure is slid by the same distance, this does not change any of the measurements of the figures involved. The result will have the same distances and angles as the original figure.

In terms of Cartesian coordinates, a translation means a shift of each of the original points (x, y) by a fixed amount in the *x* and *y* directions, to become $(x + a, y + b)$.

Another procedure that can be performed is called *reflection*. To do this, a line in the plane is specified, called the *line of reflection*. Then, each point is flipped over the line so that it is the same distance from the line but on the opposite side of it. This does not change any of the distances or angles involved, but it does reverse the order in which everything appears.

To reflect something over the *x*-axis, the points (x, y) are sent to $(x, -y)$. To reflect something over the *y*-axis, the points (x, y) are sent to the points $(-x, y)$. Flipping over other lines is not something easy to express in Cartesian coordinates. However, by drawing the figure and the line of reflection, the distance to the line and the original points can be used to find the reflected figure.

Example: Reflect this triangle with vertices (-1, 0), (2, 1), and (2, 0) over the *y*-axis. The pre-image is shown below.

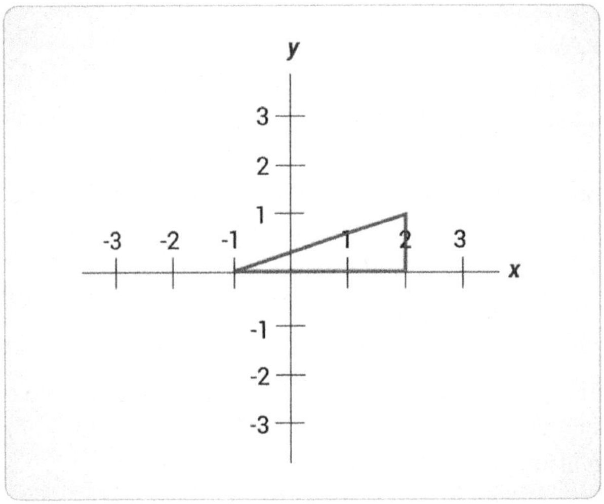

To do this, the *x*-values of the points involved are flipped to the negatives of themselves, while keeping the *y*-values the same. The image is shown here.

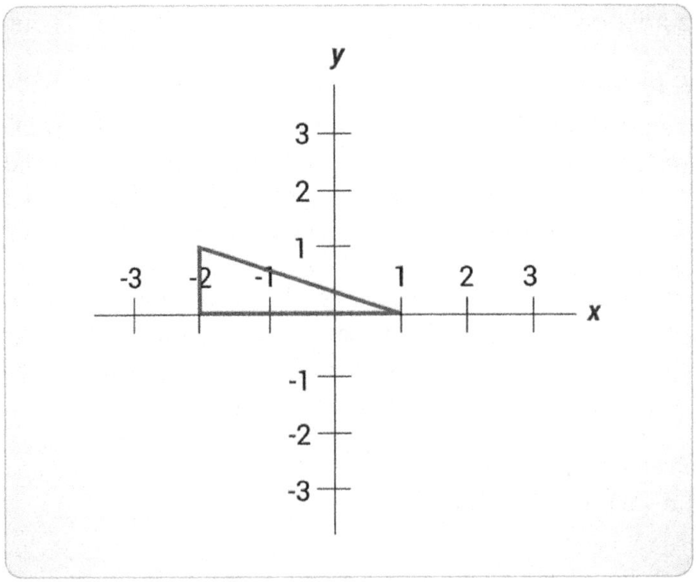

The new vertices will be (1, 0), (-2, 1), and (-2, 0).

Another procedure that does not change the distances and angles in a figure is *rotation*. In this procedure, a center point is selected, then every vertex along a circle around that point is rotated by the same angle. This procedure is also not easy to express in Cartesian coordinates, and this is not a requirement on this test. However, as with reflections, it's helpful to draw the figures and see what the result of the rotation would look like. This transformation can be performed using a compass and protractor.

Each one of these transformations can be performed on the coordinate plane without changes to the original dimensions or angles.

If two figures in the plane involve the same distances and angles, they are called *congruent figures*. In other words, two figures are congruent when they go from one form to another through reflection, rotation, and translation, or a combination of these.

Remember that rotation and translation will give back a new figure that is identical to the original figure, but reflection will give back a mirror image of it.

To recognize that a figure has undergone a rotation, one should check to see that the figure has not been changed into a mirror image, but that its orientation has changed (that is, whether the parts of the figure now form different angles with the *x*- and *y*-axes).

To recognize that a figure has undergone a translation, one should check to see that the figure has not been changed into a mirror image, and that the orientation remains the same.

To recognize that a figure has undergone a reflection, one should check to see that the new figure is a mirror image of the old figure.

Test takers should keep in mind that sometimes a combination of translations, reflections, and rotations may be performed on a figure.

Dilation

A *dilation* is a transformation that preserves angles, but not distances. It can be thought of as stretching or shrinking a figure. If a dilation makes figures larger, it is called an *enlargement*. If a dilation makes figures smaller, it is called a *reduction*. The easiest example is to dilate around the origin. In this case, the *x*- and *y*- coordinates are multiplied by a *scale factor*, k, sending points (x, y) to (kx, ky).

As an example, draw a dilation of the following triangle, whose vertices will be the points (-1, 0), (1, 0), and (1, 1).

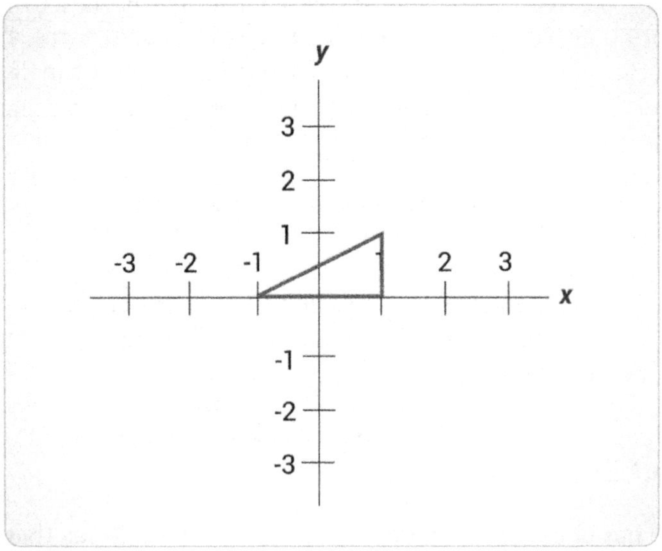

For this problem, one needs to dilate by a scale factor of 2, so the new vertices will be (-2, 0), (2, 0), and (2, 2).

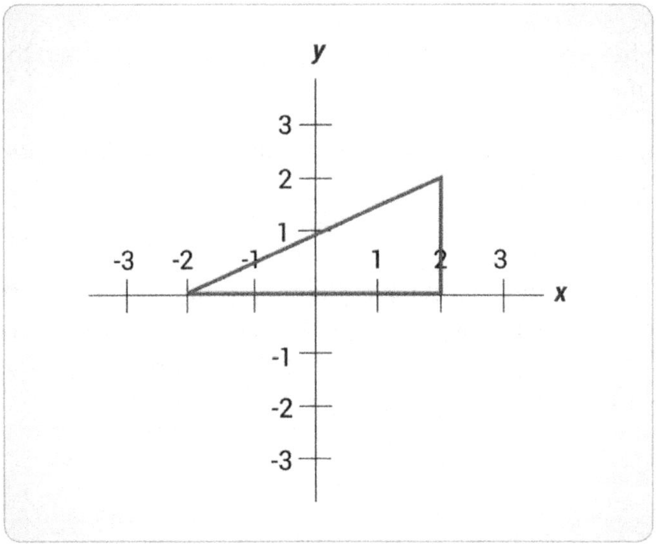

Note that after a dilation, the distances between the vertices of the figure will have changed, but the angles remain the same. The two figures that are obtained by dilation, along with possibly translation, rotation, and reflection, are all *similar* to one another. Another way to think of this is that similar figures have the same number of vertices and edges, and their angles are all the same. Similar figures have the same basic shape, but are different in size.

Symmetry

Using the types of transformations above, if an object can undergo these changes and not appear to have changed, then the figure is symmetrical. If an object can be split in half by a line and flipped over that line to lie directly on top of itself, it is said to have *line symmetry*. An example of both types of figures is seen below.

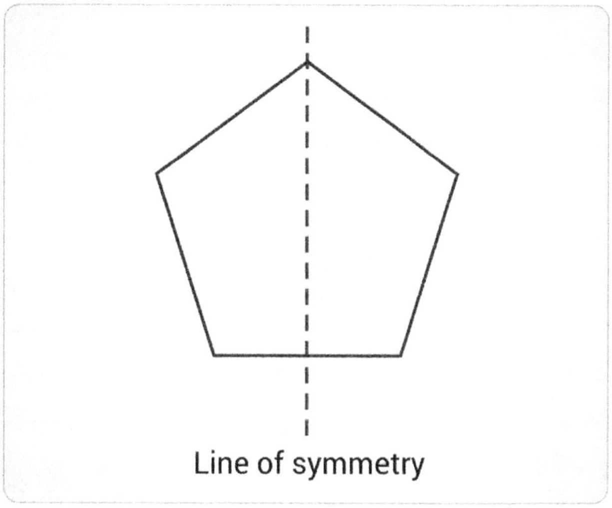

If an object can be rotated about its center to any degree smaller than 360, and it lies directly on top of itself, the object is said to have *rotational symmetry*. An example of this type of symmetry is shown below. The pentagon has an order of 5.

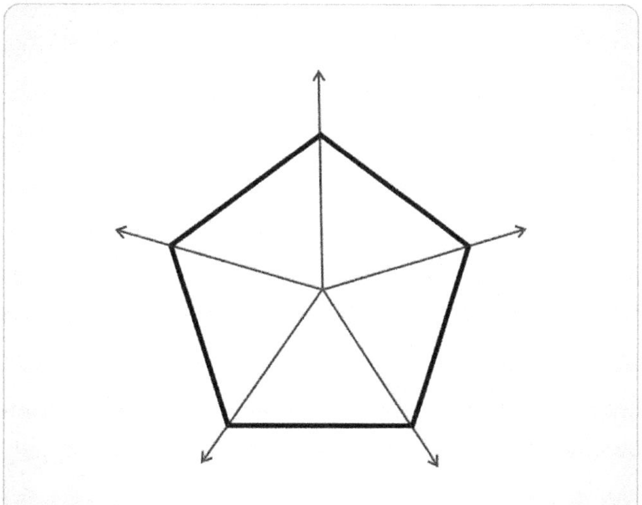

The rotational symmetry lines in the figure above can be used to find the angles formed at the center of the pentagon. Knowing that all of the angles together form a full circle, at 360 degrees, the figure can be split into five equal angles. By dividing the 360° by 5, each angle is 72°.

Given the length of one side of the figure, the perimeter of the pentagon can also be found using rotational symmetry. If one side length was 3 cm, that side length can be rotated onto each other side lengths four times. This would give a total of five side lengths equal to 3 cm. To find the perimeter, or distance around the figure, one simply needs to multiply 3 by 5. The perimeter of the figure would be 15 cm.

If a line cannot be drawn anywhere on the object to flip the figure onto itself or rotated less than or equal to 180 degrees to lay on top of itself, the object is asymmetrical. Examples of these types of figures are shown below.

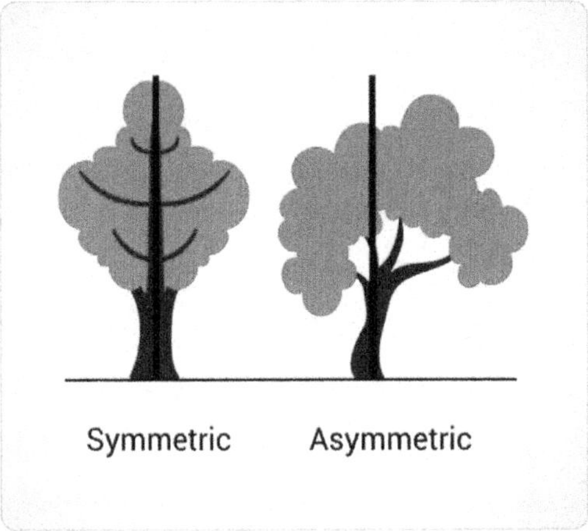

Perimeter and Area

Perimeter is the measurement of a distance around something or the sum of all sides of a polygon. Think of perimeter as the length of the boundary, like a fence. In contrast, *area* is the space occupied by a defined enclosure, like a field enclosed by a fence.

When thinking about perimeter, think about walking around the outside of something. When thinking about area, think about the amount of space or *surface area* something takes up.

Square

The perimeter of a square is measured by adding together all of the sides. Since a square has four equal sides, its perimeter can be calculated by multiplying the length of one side by 4. Thus, the formula is $P = 4 \times s$, where s equals one side. For example, the following square has side lengths of 5 meters:

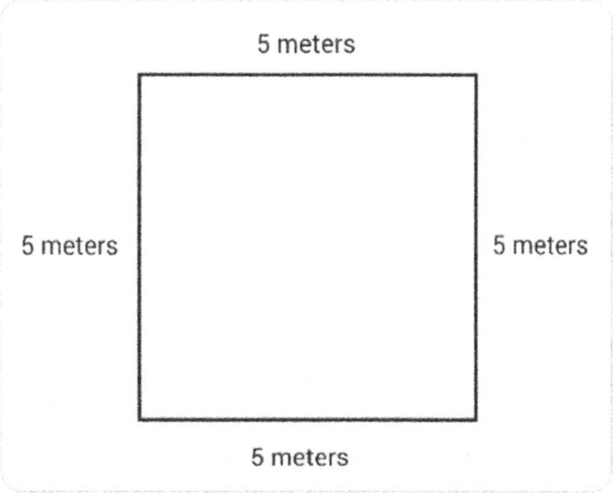

The perimeter is 20 meters because 4 times 5 is 20.

The area of a square is the length of a side squared, and the area of a rectangle is length multiplied by the width. For example, if the length of the square is 7 centimeters, then the area is 49 square centimeters. The formula for this example is $A = s^2 = 7^2 = 49$ square centimeters. An example is if the rectangle has a length of 6 inches and a width of 7 inches, then the area is 42 square inches:

$$A = lw = 6(7) = 42 \text{ square inches}$$

Rectangle

Like a square, a rectangle's perimeter is measured by adding together all of the sides. But as the sides are unequal, the formula is different. A rectangle has equal values for its lengths (long sides) and equal values for its widths (short sides), so the perimeter formula for a rectangle is:

$$P = l + l + w + w = 2l + 2w$$

l equals length
w equals width

The area is found by multiplying the length by the width, so the formula is $A = l \times w$.

For example, if the length of a rectangle is 10 inches and the width 8 inches, then the perimeter is 36 inches because:

$$P = 2l + 2w = 2(10) + 2(8)$$

$$20 + 16 = 36 \text{ inches}$$

Triangle

A triangle's perimeter is measured by adding together the three sides, so the formula is $P = a + b + c$, where a, b, and c are the values of the three sides. The area is the product of one-half the base and height so the formula is:

$$A = \frac{1}{2} \times b \times h$$

It can be simplified to:

$$A = \frac{bh}{2}$$

The base is the bottom of the triangle, and the height is the distance from the base to the peak. If a problem asks to calculate the area of a triangle, it will provide the base and height.

For example, if the base of the triangle is 2 feet and the height 4 feet, then the area is 4 square feet. The following equation shows the formula used to calculate the area of the triangle:

$$\mathrm{A} = \frac{1}{2}\mathrm{bh} = \frac{1}{2}(2)(4) = 4 \text{ square feet}$$

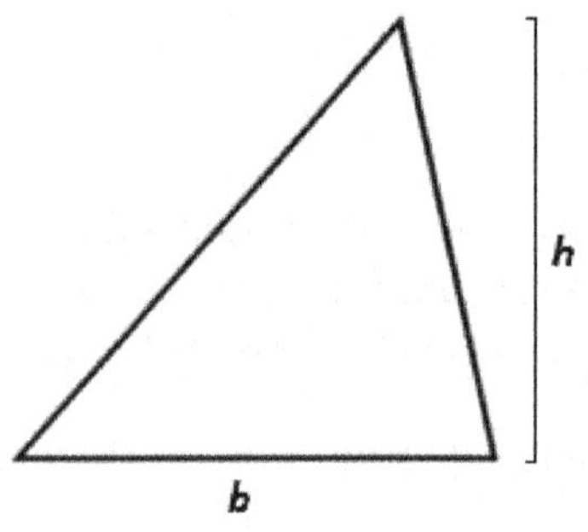

Circle

A circle's perimeter—also known as its circumference—is measured by multiplying the diameter by π.

Diameter is the straight line measured from one end to the direct opposite end of the circle.

π is referred to as pi and is equal to 3.14 (with rounding).

So the formula is $\pi \times d$.

This is sometimes expressed by the formula $C = 2 \times \pi \times r$, where r is the radius of the circle. These formulas are equivalent, as the radius equals half of the diameter.

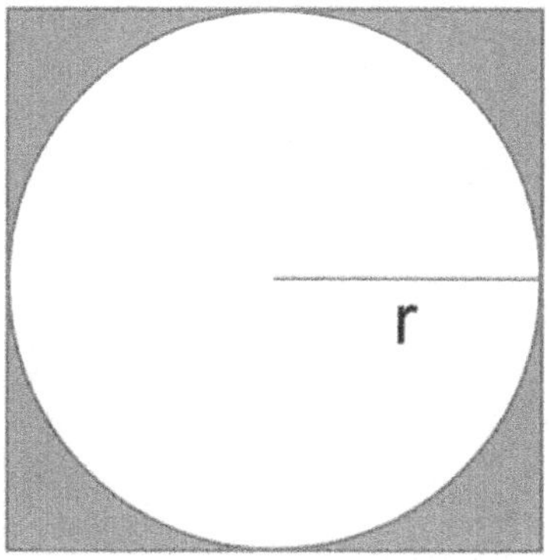

The area of a circle is calculated through the formula $A = \pi \times r^2$. The test will indicate either to leave the answer with π attached or to calculate to the nearest decimal place, which means multiplying by 3.14 for π.

Arc

The *arc of a circle* is the distance between two points on the circle. The length of the arc of a circle in terms of *degrees* is easily determined if the value of the central angle is known. The length of the arc is simply the value of the central angle. In this example, the length of the arc of the circle in degrees is 75°.

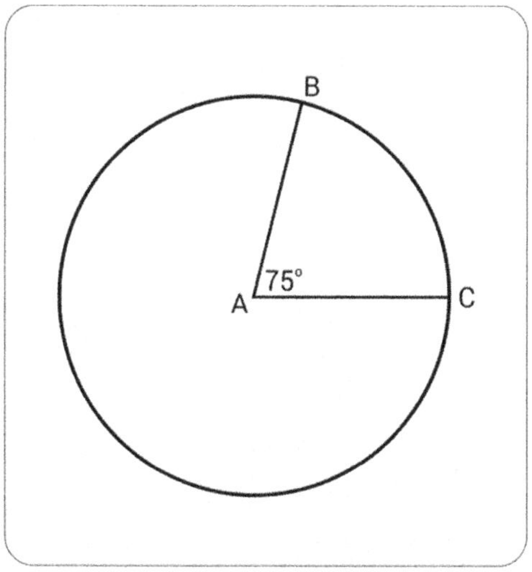

To determine the length of the arc of a circle in *distance*, the student will need to know the values for both the central angle and the radius. This formula is:

$$\frac{central\ angle}{360°} = \frac{arc\ length}{2\pi r}$$

The equation is simplified by cross-multiplying to solve for the arc length.

In the following example, the student should substitute the values of the central angle (75°) and the radius (10 inches) into the equation above to solve for the arc length.

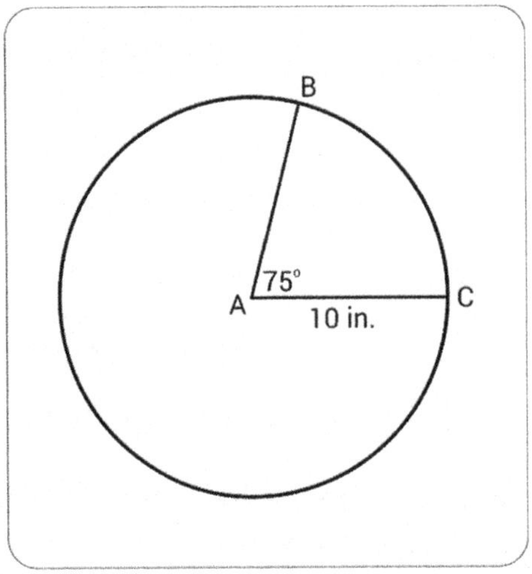

$$\frac{75°}{360°} = \frac{\text{arc length}}{2(3.14)(10\text{in.})}$$

To solve the equation, first cross-multiply: 4710 = 360(arc length). Next, divide each side of the equation by 360. The result of the formula is that the arc length is 13.1 (rounded).

Irregular Shapes

The perimeter of an irregular polygon is found by adding the lengths of all of the sides. In cases where all of the sides are given, this will be very straightforward, as it will simply involve finding the sum of the provided lengths. Other times, a side length may be missing and must be determined before the perimeter can be calculated. Consider the example below:

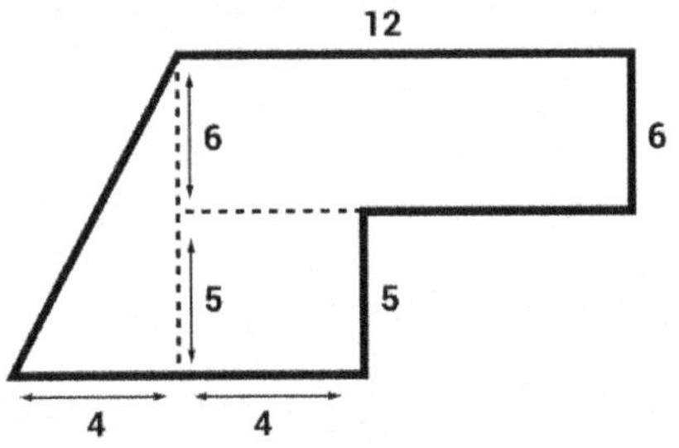

All of the side lengths are provided except for the angled side on the left. Test takers should notice that this is the hypotenuse of a right triangle. The other two sides of the triangle are provided (the base is 4 and the height is 6 + 5 = 11). The Pythagorean Theorem can be used to find the length of the hypotenuse, remembering that $a^2 + b^2 = c^2$.

Substituting the side values provided yields $(4)^2 + (11)^2 = c^2$.

Therefore, $c = \sqrt{16 + 121} = 11.7$

Finally, the perimeter can be found by adding this new side length with the other provided lengths to get the total length around the figure: 4+4+5+8+6+12+11.7=50.7. Although units are not provided in this figure, remember that reporting units with a measurement is important.

The area of an irregular polygon is found by decomposing, or breaking apart, the figure into smaller shapes. When the area of the smaller shapes is determined, these areas are added together to produce the total area of the area of the original figure. Consider the same example provided before:

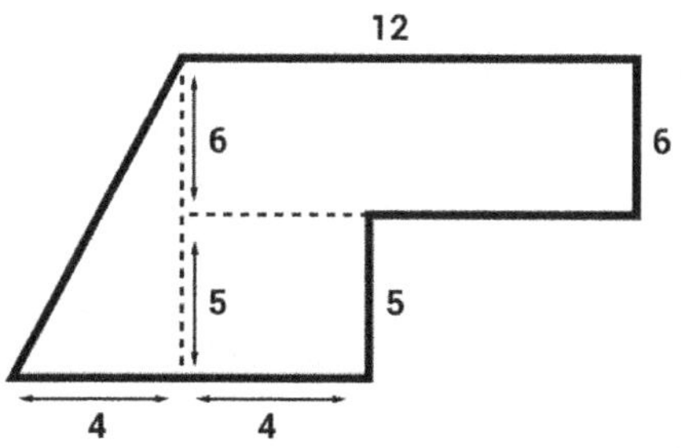

The irregular polygon is decomposed into two rectangles and a triangle. The area of the large rectangles ($A = l \times w \rightarrow A = 12 \times 6$) is 72 square units. The area of the small rectangle is 20 square units ($A = 4 \times 5$). The area of the triangle ($A = \frac{1}{2} \times b \times h \rightarrow A = \frac{1}{2} \times 4 \times 11$) is 22 square units. The sum of the areas of these figures produces the total area of the original polygon:

$$A = 72 + 20 + 22 \rightarrow A = 114 \text{ square units}$$

Volumes and Surface Areas

Geometry in three dimensions is similar to geometry in two dimensions. The main new feature is that three points now define a unique *plane* that passes through each of them. Three-dimensional objects can be made by putting together two-dimensional figures in different surfaces. Below, some of the possible three-dimensional figures will be provided, along with formulas for their volumes and surface areas.

A rectangular prism is a box whose sides are all rectangles meeting at 90° angles. Such a box has three dimensions: length, width, and height. If the length is *x*, the width is *y*, and the height is *z*, then the volume is given by $V = xyz$.

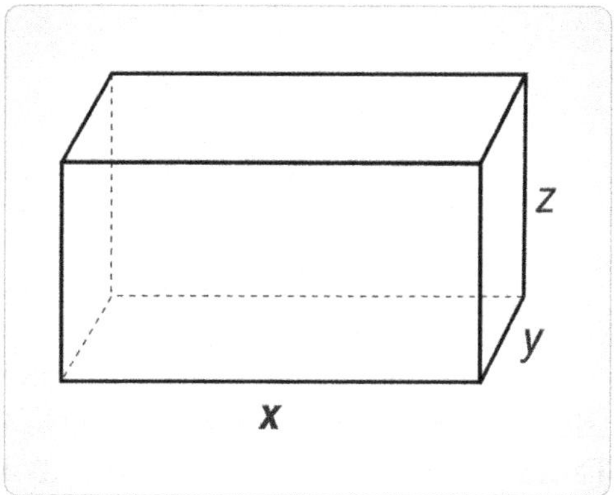

The surface area will be given by computing the surface area of each rectangle and adding them together. There are a total of six rectangles. Two of them have sides of length *x* and *y*, two have sides of length *y* and *z*, and two have sides of length *x* and *z*. Therefore, the total surface area will be given by:

$$SA = 2xy + 2yz + 2xz$$

A *rectangular pyramid* is a figure with a rectangular base and four triangular sides that meet at a single vertex. If the rectangle has sides of lengths *x* and *y*, then the volume will be given by $V = \frac{1}{3}xyh$.

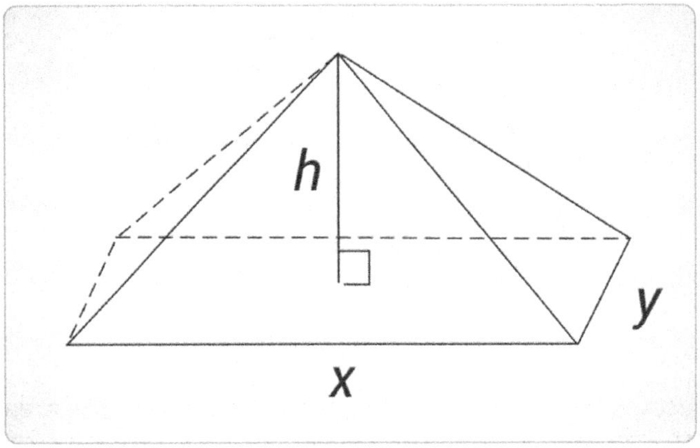

To find the surface area, the dimensions of each triangle need to be known. However, these dimensions can differ depending on the problem in question. Therefore, there is no general formula for calculating total surface area.

A *sphere* is a set of points all of which are equidistant from some central point. It is like a circle, but in three dimensions. The volume of a sphere of radius *r* is given by $V = \frac{4}{3}\pi r^3$. The surface area is given by $A = 4\pi r^2$.

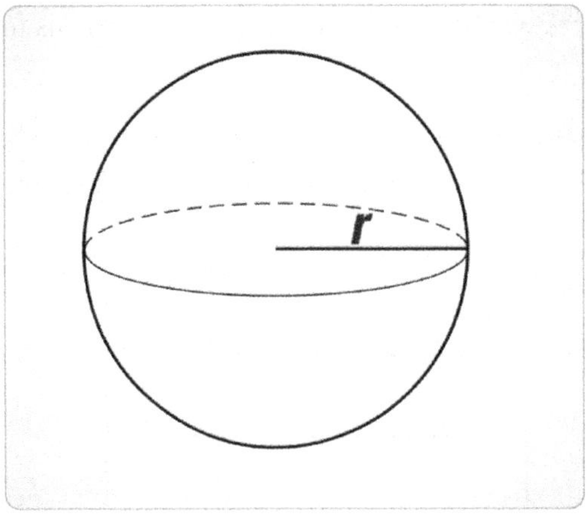

The Pythagorean Theorem

The Pythagorean theorem is an important concept in geometry. It states that for right triangles, the sum of the squares of the two shorter sides will be equal to the square of the longest side (also called the *hypotenuse*). The longest side will always be the side opposite to the 90° angle. If this side is called *c*, and the other two sides are a and b, then the Pythagorean theorem states that $c^2 = a^2 + b^2$. Since lengths are always positive, this also can be written as $c = \sqrt{a^2 + b^2}$. A diagram to show the parts of a triangle using the Pythagorean theorem is below.

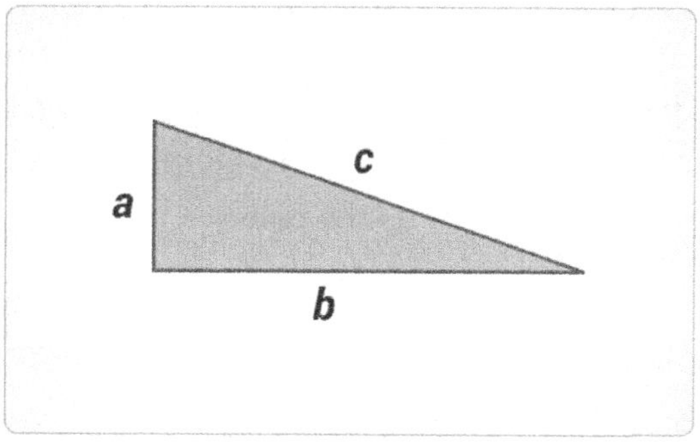

As an example of the theorem, suppose that Shirley has a rectangular field that is 5 feet wide and 12 feet long, and she wants to split it in half using a fence that goes from one corner to the opposite corner. How long will this fence need to be? To figure this out, note that this makes the field into two right triangles, whose hypotenuse will be the fence dividing it in half. Therefore, the fence length will be given by:

$$\sqrt{5^2 + 12^2} = \sqrt{169} = 13 \text{ feet long}$$

Similar Figures and Proportions

Sometimes, two figures are similar, meaning they have the same basic shape and the same interior angles, but they have different dimensions. If the ratio of two corresponding sides is known, then that ratio, or scale factor, holds true for all of the dimensions of the new figure.

Here is an example of applying this principle. Suppose that Lara is 5 feet tall and is standing 30 feet from the base of a light pole, and her shadow is 6 feet long. How high is the light on the pole? To figure this out, it helps to make a sketch of the situation:

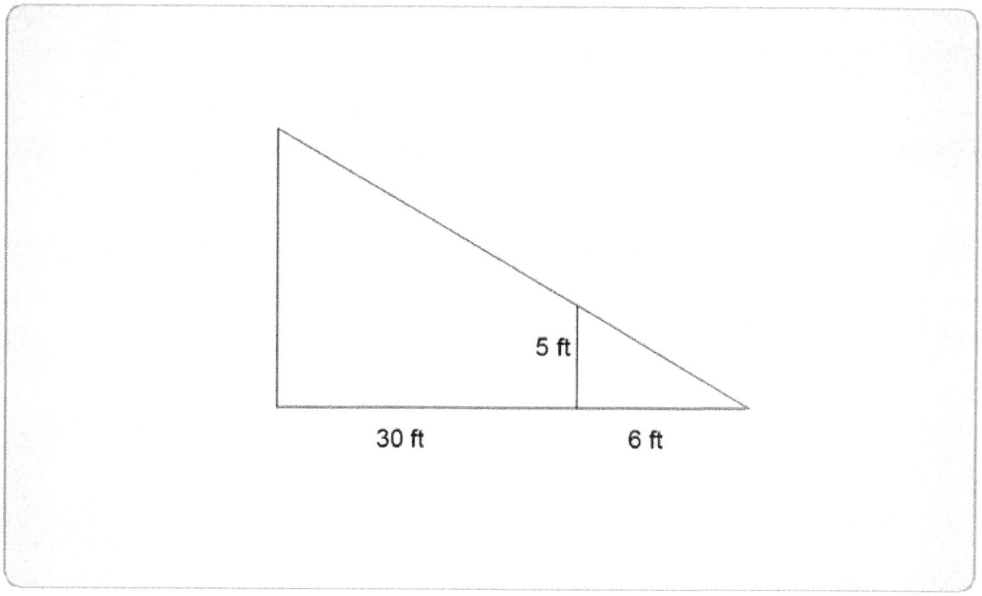

The light pole is the left side of the triangle. Lara is the 5-foot vertical line. Notice that there are two right triangles here, and that they have all the same angles as one another. Therefore, they form similar triangles. So, the ratio of proportionality between them needs to be found.

The bases of these triangles are known. The small triangle, formed by Lara and her shadow, has a base of 6 feet. The large triangle, formed by the light pole along with the line from the base of the pole out to the end of Lara's shadow is $30 + 6 = 36$ feet long. So, the ratio of the big triangle to the little triangle will be $\frac{36}{6} = 6$. The height of the little triangle is 5 feet. Therefore, the height of the big triangle will be $6 \cdot 5 = 30$ feet, meaning that the light is 30 feet up the pole.

Notice that the perimeter of a figure changes by the ratio of proportionality between two similar figures, but the area changes by the *square* of the ratio. This is because if the length of one side is doubled, the area is quadrupled.

As an example, suppose two rectangles are similar, but the edges of the second rectangle are three times longer than the edges of the first rectangle. The area of the first rectangle is 10 square inches. How much more area does the second rectangle have than the first?

To answer this, note that the area of the second rectangle is $3^2 = 9$ times the area of the first rectangle, which is 10 square inches. Therefore, the area of the second rectangle is going to be $9 \cdot 10 = 90$ square inches. This means it has $90 - 10 = 80$ square inches more area than the first rectangle.

As a second example, suppose *X* and *Y* are similar right triangles. The hypotenuse of *X* is 4 inches. The area of *Y* is $\frac{1}{4}$ the area of *X*. What is the hypotenuse of *Y*?

The area has changed by a factor of $\frac{1}{4}$. The area changes by a factor that is the *square* of the ratio of changes in lengths, so the ratio of the lengths is the square root of the ratio of areas. That means that the ratio of lengths must be is $\sqrt{\frac{1}{4}} = \frac{1}{2}$, and the hypotenuse of *Y* must be $\frac{1}{2} \cdot 4 = 2$ inches.

Volumes between similar solids change like the cube of the change in the lengths of their edges. Likewise, if the ratio of the volumes between similar solids is known, the ratio between their lengths is known by finding the cube root of the ratio of their volumes.

For example, suppose there are two similar rectangular pyramids *X* and *Y*. The base of *X* is 1 inch by 2 inches, and the volume of *X* is 8 inches. The volume of *Y* is 64 inches. What are the dimensions of the base of *Y*?

To answer this the ratio of the volume of *Y* to the volume of *X* is found first. This will be given by $\frac{64}{8} = 8$. Now the ratio of lengths is the cube root of the ratio of volumes, or $\sqrt[3]{8} = 2$. So, the dimensions of the base of *Y* must be 2 inches by 4 inches.

Practice Questions

1. Which of the following numbers has the greatest value?
    a. 1.4378
    b. 1.07548
    c. 1.43592
    d. 0.89409
    e. 1.43688

2. The value of 6×12 is the same as which of the following?
    a. $2 \times 4 \times 4 \times 2$
    b. $7 \times 4 \times 3$
    c. $6 \times 6 \times 3$
    d. $3 \times 3 \times 4 \times 2$
    e. $3 \times 4 \times 6 \times 2$

3. The area of a given rectangle is 24 centimeters. If the measure of each side is multiplied by 3, what is the area of the new figure?
    a. 48 cm^2
    b. 72 cm^2
    c. 216 cm^2
    d. 13,824 cm^2
    e. 224 cm^2

4. After a 20% sale discount, Frank purchased a new refrigerator for $850. How much did he save from the original price?
    a. $170
    b. $212.50
    c. $105.75
    d. $200
    e. $150

5. A student gets an 85% on a test with 20 questions. How many answers did the student solve correctly?
    a. 16
    b. 15
    c. 18
    d. 19
    e. 17

6. Alan currently weighs 200 pounds, but he wants to lose weight to get down to 175 pounds. What is this difference in kilograms? (1 pound is approximately equal to 0.45 kilograms.)
    a. 9 kg
    b. 11.25 kg
    c. 78.75 kg
    d. 90 kg
    e. 25 kg

7. Johnny earns \$2334.50 from his job each month. He pays \$1437 for monthly expenses. Johnny is planning a vacation in 3 months' time that he estimates will cost \$1750 total. How much will Johnny have left over from three months' of saving once he pays for his vacation?

a. \$948.50
b. \$584.50
c. \$852.50
d. \$942.50
e. \$952.50

8. What is $\frac{420}{98}$ rounded to the nearest integer?

a. 3
b. 4
c. 5
d. 6
e. 7

9. Being as specific as possible, how is the number −4 classified?

a. Real, rational, integer, whole, natural
b. Real, rational, integer, natural
c. Real, rational, integer
d. Real, irrational, whole
e. Real, irrational, complex

10. Which of the following is the correct simplification of: $(7n + 3n^3 + 3) + (8n + 5n^3 + 2n^4)$?

a. $9n^4 + 15n - 2$
b. $2n^4 + 5n^3 + 15n - 2$
c. $9n^4 + 8n^3 + 15n$
d. $2n^3 + 8n^2 + 15n + 3$
e. $2n^4 + 8n^3 + 15n + 3$

11. What is the solution to $(\sqrt{36} \times \sqrt{16}) - 3^2$?

a. 30
b. 21
c. 15
d. 13
e. 16

12. In Jim's school, there are 3 girls for every 2 boys. There are 650 students in total. Using this information, how many students are girls?

a. 260
b. 130
c. 65
d. 390
e. 225

13. Five of six numbers have a sum of 25. The average of all six numbers is 6. What is the sixth number?
a. 8
b. 12
c. 13
d. 10
e. 11

14. What is the value of the Roman numeral CCXLVII equal?
a. 377
b. 301
c. 251
d. 247
e. 3107

15. Which of the following correctly arranges the numbers from least to greatest value?
$0.85, \frac{4}{5}, \frac{2}{3}, \frac{91}{100}$

a. $0.85, \frac{4}{5}, \frac{2}{3}, \frac{91}{100}$
b. $\frac{4}{5}, 0.85, \frac{91}{100}, \frac{2}{3}$
c. $\frac{2}{3}, \frac{4}{5}, 0.85, \frac{91}{100}$
d. $0.85, \frac{91}{100}, \frac{4}{5}, \frac{2}{3}$
e. $\frac{4}{5}, \frac{2}{3}, 0.85, \frac{91}{100}$

16. Keith's bakery had 252 customers go through its doors last week. This week, that number increased to 378. How is this increased expressed as a percentage?
a. 26%
b. 50%
c. 35%
d. 12%
e. 28%

17. A company invests $50,000 in a building where they can produce saws. If the cost of producing one saw is $40, then which function expresses the amount of money the company pays? The variable *y* is the money paid and *x* is the number of saws produced.
a. $y = 50{,}000x + 40$
b. $y + 40 = x - 50{,}000$
c. $y = 40x - 50{,}000$
d. $y = 40x + 50{,}000$
e. $y = 4(x + 50{,}000)$

18. Four people split a bill. The first person pays for $\frac{1}{5}$, the second person pays for $\frac{1}{4}$, and the third person pays for $\frac{1}{3}$. What fraction of the bill does the fourth person pay?

a. $\frac{13}{60}$

b. $\frac{47}{60}$

c. $\frac{1}{4}$

d. $\frac{4}{15}$

e. $\frac{1}{2}$

19. Which of the following inequalities is equivalent to $3 - \frac{1}{2}x \geq 2$?

a. $x \geq 2$
b. $x \leq 2$
c. $x \geq 1$
d. $x \leq 1$
e. $x \leq 3$

20. A closet is filled with red, blue, and green shirts. If $\frac{1}{3}$ of the shirts are green and $\frac{2}{5}$ are red, what fraction of the shirts are blue?

a. $\frac{4}{15}$

b. $\frac{1}{5}$

c. $\frac{7}{15}$

d. $\frac{1}{2}$

e. $\frac{2}{3}$

21. Shawna buys $2\frac{1}{2}$ gallons of paint. If she uses $\frac{1}{3}$ of it on the first day, how much does she have left?

a. $1\frac{5}{6}$ gallons

b. $1\frac{1}{2}$ gallons

c. $1\frac{2}{3}$gallons

d. 2 gallons

e. $1\frac{3}{4}$ gallons

22. Jessica buys 10 cans of paint. Red paint costs \$1 per can and blue paint costs \$2 per can. In total, she spends \$16. How many red cans did she buy?

a. 2
b. 3
c. 4
d. 5
e. 6

23. Which of the following is an equation for the line passing through the origin and the point $(2, 1)$?

a. $y = 2x$
b. $y = \frac{1}{2}x$
c. $y = x - 2$
d. $2y = x + 1$
e. $y = -\frac{1}{2}x$

24. Which of the following is equivalent to the value of the digit 3 in the number 792.134?

a. 3×10
b. 3×100
c. $\frac{3}{10}$
d. $\frac{3}{100}$
e. 3×0.1

25. Which of the following represents the expression for three times the sum of twice a number and one minus six?

a. $2x + 1 - 6$
b. $3x + 1 - 6$
c. $3(x + 1) - 6$
d. $3(2x + 1) - 6$
e. $2(3x + 1) - 6$

26. How is the number 847.89632 written if it is rounded to the nearest hundredth?

a. 847.90
b. 900
c. 847.89
d. 847.896
e. 847.895

27. The perimeter of a 6-sided polygon is 56 cm. Three of the sides have a length of 9 cm each. Two other sides are 8 cm each. What is the length of the missing side?

a. 11 cm
b. 12 cm
c. 13 cm
d. 10 cm
e. 9 cm

28. If Danny takes 48 minutes to walk 3 miles, how long should it take him to walk 5 miles if he maintains the same speed?
    a. 32 min
    b. 64 min
    c. 80 min
    d. 96 min
    e. 78 min

29. If Sarah reads at an average rate of 21 pages in four nights, how long will it take her to read 140 pages?
    a. 6 nights
    b. 26 nights
    c. 8 nights
    d. 27 nights
    e. 7 nights

30. The phone bill is calculated each month using the equation $c = 50g + 75$. The cost of the phone bill per month is represented by c, and g represents the number of gigabytes of data used that month. What is the value and interpretation of the slope of this equation?
    a. 75 dollars per day
    b. 75 gigabytes per day
    c. 50 dollars per day
    d. 50 dollars per gigabyte
    e. 125 dollars per gigabyte

31. What are the polynomial roots of $x^2 + x - 2$?
    a. 1 and −2
    b. −1 and 2
    c. 2 and -2
    d. 9 and 13
    e. −1 and −2

32. A square has a side length of 4 inches. A triangle has a base of 2 inches and a height of 8 inches. What is the total area of the square and triangle?
    a. 24 square inches
    b. 28 square inches
    c. 32 square inches
    d. 36 square inches
    e. 40 square inches

33. A triangle is to have a base one-third as long as its height. Its area must be 6 square feet. How long will its base be?
    a. 1 foot
    b. 1.5 feet
    c. 2 feet
    d. 2.5 feet
    e. 3 feet

34. For the following similar triangles, what are the values of x and y (rounded to one decimal place)?

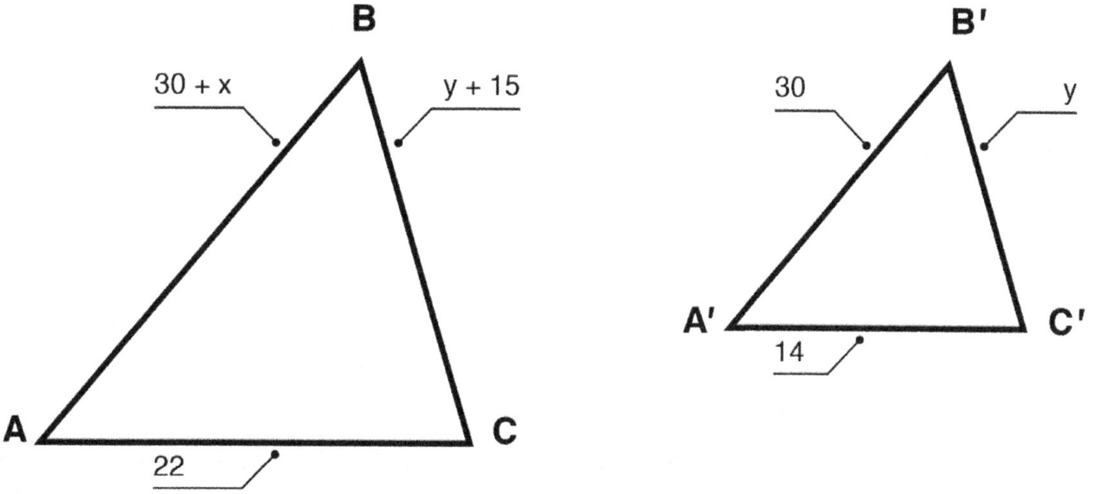

a. $x = 16.5, y = 25.1$
b. $x = 19.5, y = 24.1$
c. $x = 17.1, y = 26.3$
d. $x = 26.3, y = 17.1$
e. $x = 25.1, y = 17.3$

35. What is the value of $x^2 - 2xy + 2y^2$ when $x = 2, y = 3$?
a. 8
b. 10
c. 12
d. 14
e. 4

36. What is the y-intercept for $y = x^2 + 3x - 4$?
a. $y = 1$
b. $y = -3$
c. $y = 3$
d. $y = 4$
e. $y = -4$

37. If the volume of a sphere is 288π cubic meters, what are the radius and surface area of the same sphere?
a. Radius 6 meters and surface area 144π square meters
b. Radius 36 meters and surface area 144π square meters
c. Radius 6 meters and surface area 12π square meters
d. Radius 36 meters and surface area 12π square meters
e. Radius 6 meters and surface area 36π square meters

38. A rectangle has a length that is 5 feet longer than three times its width. If the perimeter is 90 feet, what is the length in feet?

a. 10
b. 20
c. 25
d. 35
e. 15

39. What is the y-intercept of $y = x^{5/3} + (x - 3)(x + 1)$?

a. 3.5
b. 7.6
c. −3
d. −15.1
e. −5

40. A farmer owns two (non-adjacent) square plots of land, which he wishes to fence. The area of one is 1000 square feet, while the area of the other is 10 square feet. How much fencing does he need, in feet?

a. 44
b. $40\sqrt{10}$
c. $440\sqrt{10}$
d. $40 + 4\sqrt{10}$
e. $44\sqrt{10}$

Answer Explanations

1. A: Compare each numeral after the decimal point to figure out which overall number is greatest. Both Choices *A* (1.43785) and *C* (1.43592) have the same tenths (4) and hundredths (3). However, the thousandths is greater in answer *A* (7), so *A* has the greatest value overall.

2. D: By grouping the four numbers in the answer into factors of the two numbers of the question (6 and 12), it can be determined that $(3 \times 2)(4 \times 3) = 6 \times 12$. Alternatively, each of the answer choices could be prime factored or multiplied out and compared to the original value. 6×12 has a value of 72 and a prime factorization of $2^3 \times 3^2$. The answer choices respectively have values of 64, 84, 108, 72, and 144 and prime factorizations of 2^6, $2^2 \times 3 \times 7$, $2^2 \times 3^3$, $2^3 \times 3^2$, and $2 \times 2^3 \times 3^2$ so Choice *D* is the correct answer.

3. C: Because area is a two-dimensional measurement, the dimensions are multiplied by a factor that is squared to determine the scale of the corresponding areas. The dimensions of the rectangle are multiplied by a scale of 3. Therefore, the area is multiplied by a scale of 3^2 or 9, so $24cm \times 9 = 216cm$.

4. B: Since $850 is the price *after* a 20% discount, $850 represents 80% of the original price. To determine the original price, set up a proportion with the ratio of the sale price (850) to original price (unknown) equal to the ratio of sale percentage:

$$\frac{850}{x} = \frac{80}{100}$$

(where *x* represents the unknown original price)

To solve a proportion, cross-multiply the numerators and denominators and set the products equal to each other:

$$(850)(100) = (80)(x)$$

Multiplying each side results in the equation 85,000 = 80x.

To solve for *x*, divide both sides by 80: $\frac{85{,}000}{80} = \frac{80x}{80}$, which yields *x* = 1062.5. Remember that *x* represents the original price. Subtracting the sale price from the original price ($1062.50-$850) reveals that Frank saved $212.50.

5. E: 85% of a number means that number should be multiplied by 0.85: $0.85 \times 20 = \frac{85}{100} \times \frac{20}{1}$, which can be simplified to $\frac{17}{20} \times \frac{20}{1} = 17$.

6. B: Using the conversion rate, multiply the projected weight loss of 25 lb. by 0.45 $\frac{kg}{lb}$ to get the desired amount in kilograms (11.25 kg).

7. D: First, subtract $1437 from $2334.50 to find Johnny's monthly savings; this equals $897.50. Then, multiply this amount by 3 to find out how much he will have (in three months) before he pays for his vacation: this equals $2692.50. Finally, subtract the cost of the vacation ($1750) from this sum to find how much Johnny will have left: $942.50.

8. B: Dividing by 98 can be approximated by dividing by 100, which would mean shifting the decimal point of the numerator to the left two places. The result is 4.2, which rounds to 4.

9. C: The number −4 is classified as a real number because it exists and is not imaginary. It is rational because it does not have a decimal that never ends. It is an integer because it does not have a fractional component. The next classification would be whole numbers, for which −4 does not qualify because it is negative. Although −4 could technically be considered a complex number because complex numbers can have either the real or imaginary part equal zero and still be considered a complex number, Choices *D* and *E* are wrong because −4 is not considered an irrational number because it does not have a never-ending decimal component.

10. E: The expression is simplified by collecting like terms. Terms with the same variable and exponent are like terms, and their coefficients can be added.

11. C: Follow the *order of operations* in order to solve this problem. Solve the parentheses first, and then follow the remainder as usual.

$$(6 \times 4) - 9$$

This equals $24 - 9$, or 15, which is Choice *C*.

12. D: Three girls for every two boys can be expressed as a ratio: 3:2. This can be visualized as splitting the school into 5 groups: 3 girl groups and 2 boy groups. The number of students that are in each group can be found by dividing the total number of students by 5:

650 divided by 5 equals 1 part, or 130 students per group

To find the total number of girls, multiply the number of students per group (130) by the number of girl groups in the school (3). This equals 390, which is Choice *D*.

13. E: If the average of all six numbers is 6, that means $\frac{a+b+c+d+e+x}{6} = 6$. The sum of the first five numbers is 25, so this equation can be simplified to $\frac{25+x}{6} = 6$. Multiplying both sides by 6 gives $25 + x = 36$, and x—or the sixth number—must equal 11.

14. D: In Roman numerals, C is 100, L is 50, X is 10, V is 5, and I is 1. To prevent four identical letters in a row, a lesser number is put before a larger one, and the lesser number is subtracted from the larger one. In this case, an X (10) is put before an L (50), which equals 40. Written in Arabic form, the numerals are 200 + 40 +5 +2 = 247.

15. C: The first step is to depict each number using decimals. $\frac{91}{100}$ = 0.91

Dividing the numerator by denominator of $\frac{4}{5}$ to convert it to a decimal yields 0.80, while $\frac{2}{3}$ becomes 0.66 recurring. Rearrange each expression in ascending order, as found in answer *C*.

16. B: First, calculate the difference between the larger value and the smaller value.

378 – 252 = 126

To calculate this difference as a percentage of the original value, and thus calculate the percentage *increase*, divide 126 by 252, then multiply by 100 to reach the percentage = 50%, Choice *B*.

17. D: For manufacturing costs, there is a linear relationship between the cost to the company and the number produced, with a *y*-intercept given by the base cost of acquiring the means of production, and a slope given by the cost to produce one unit. In this case, that base cost is $50,000, while the cost per unit is $40. So, $y = 40x + 50{,}000$.

18. A: To find the fraction of the bill that the first three people pay, the fractions need to be added, which means finding common denominator. The common denominator will be 60.

$$\frac{1}{5} + \frac{1}{4} + \frac{1}{3} = \frac{12}{60} + \frac{15}{60} + \frac{20}{60} = \frac{47}{60}$$

The remainder of the bill is $1 - \frac{47}{60} = \frac{60}{60} - \frac{47}{60} = \frac{13}{60}$.

19. B: To simplify this inequality, subtract 3 from both sides to get $-\frac{1}{2}x \geq -1$. Then, multiply both sides by -2 (remembering that this flips the direction of the inequality) to get $x \leq 2$.

20. A: The total fraction taken up by green and red shirts will be $\frac{1}{3} + \frac{2}{5} = \frac{5}{15} + \frac{6}{15} = \frac{11}{15}$.

The remaining fraction is $1 - \frac{11}{15} = \frac{15}{15} - \frac{11}{15} = \frac{4}{15}$.

21. C: If she has used one-third of the paint, she has two-thirds remaining. $2\frac{1}{2}$ gallons is the same as $\frac{5}{2}$ gallons. The calculation is:

$$\frac{2}{3} \times \frac{5}{2} = \frac{5}{3} = 1\frac{2}{3} \text{ gallons}$$

22. C: We are trying to find x, the number of red cans. The equation can be set up like this:

$$x + 2(10 - x) = 16$$

The left x is actually multiplied by $1, the price per red can. Since we know Jessica bought 10 total cans, $10 - x$ is the number blue cans that she bought. We multiply the number of blue cans by $2, the price per blue can.

That should all equal $16, the total amount of money that Jessica spent. Working that out gives us:

$$x + 20 - 2x = 16$$

$$20 - x = 16$$

$$x = 4$$

23. B: The slope will be given by $\frac{1-0}{2-0} = \frac{1}{2}$. The *y*-intercept will be 0, since it passes through the origin. Using slope-intercept form, the equation for this line is $y = \frac{1}{2}x$.

24. D: Each digit to the left of the decimal point represents a higher multiple of 10 and each digit to the right of the decimal point represents a quotient of a higher multiple of 10 for the divisor. The first digit to the right of the decimal point is equal to the value $\div$ 10. The second digit to the right of the decimal point is equal to the value $\div (10 \times 10)$, or the value $\div$ 100, so the answer is $\frac{3}{100}$.

25. D: The expression is three times the sum of twice a number and 1, which is $3(2x + 1)$. Then, 6 is subtracted from this expression.

26. A: The hundredths place value is located two digits to the right of the decimal point (the digit 9). The digit to the right of the place value is examined to decide whether to round up or keep the digit. In this case, the digit 6 is 5 or greater, so the hundredths place is rounded up. When rounding up, if the digit to be increased is a 9, the digit to its left is increased by one and the digit in the desired place value becomes a zero. Therefore, the number is rounded to 847.90.

27. C: Perimeter is found by calculating the sum of all sides of the polygon. $9 + 9 + 9 + 8 + 8 + s = 56$, where *s* is the missing side length. Therefore, 43 plus the missing side length is equal to 56. The missing side length is 13 cm.

28. C: To solve the problem, a proportion is written consisting of ratios comparing distance and time. One way to set up the proportion is: $\frac{3}{48} = \frac{5}{x}\left(\frac{distance}{time} = \frac{distance}{time}\right)$, where *x* represents the unknown value of time. To solve a proportion, the ratios are cross-multiplied:

$$(3)(x) = (5)(48) \rightarrow 3x = 240$$

The equation is solved by isolating the variable, or dividing by 3 on both sides, to produce $x = 80$.

29. D: This problem can be solved by setting up a proportion involving the given information and the unknown value. The proportion is:

$$\frac{21\ pages}{4\ nights} = \frac{140\ pages}{x\ nights}$$

Solving the proportion by cross-multiplying, the equation becomes $21x = 4 * 140$, where $x = 26.67$. Since it is not an exact number of nights, the answer is rounded up to 27 nights. Twenty-six nights would not give Sarah enough time.

30. D: The slope from this equation is 50, and it is interpreted as the cost per gigabyte used. Since the g-value represents number of gigabytes and the equation is set equal to the cost in dollars, the slope relates these two values. For every gigabyte used on the phone, the bill increases 50 dollars.

31. A: Finding the roots means finding the values of *x* when *y* is zero. The quadratic formula could be used, but in this case, it is possible to factor by hand, since the numbers -1 and 2 add to 1 and multiply to -2. So, factor $x^2 + x - 2 = (x - 1)(x + 2) = 0$, then set each factor equal to zero. Solving for each value gives the values *x* = 1 and *x* = -2.

32. A: The area of the square is the square of its side length, so $4^2 = 16$ square inches. The area of a triangle is half the base times the height, so $\frac{1}{2} \times 2 \times 8 = 8$ square inches. The total is $16 + 8 = 24$ square inches.

33. C: The formula for the area of a triangle with base b and height h is $\frac{1}{2}bh$, where the base is one-third the height, or $b = \frac{1}{3}h$ or equivalently $h = 3b$.

Using the formula for a triangle, this becomes:

$$\frac{1}{2}b(3b) = \frac{3}{2}b^2$$

Now, this has to be equal to 6. So $\frac{3}{2}b^2 = 6$, $b^2 = 4$, and $b = \pm 2$. However, lengths are positive, so the base must be 2 feet long.

34. C: Because the triangles are similar, the lengths of the corresponding sides are proportional. Therefore:

$$\frac{30 + x}{30} = \frac{22}{14} = \frac{y + 15}{y}$$

This results in the equation $14(30 + x) = 22 \cdot 30$ which, when solved, gives $x = 17.1$. The proportion also results in the equation $14(y + 15) = 22y$ which, when solved, gives $y = 26.3$.

35. B: Start with the original equation: x- 2xy + 2y, then replace each instance of x with a 2, and each instance of y with a 3 to get:

$$2^2 - 2 \cdot 2 \cdot 3 + 2 \cdot 3^2 = 4 - 12 + 18 = 10$$

36. E: The y-intercept of an equation is found where the x-value is zero. Plugging zero into the equation for x, the first two terms cancel out, leaving -4.

37. A: Because the volume of the given sphere is 288π cubic meters, this means $^4/_3\,\pi r^3 = 288\pi$. This equation is solved for r to obtain a radius of 6 meters. The formula for the surface area of a sphere is $4\pi r^2$, so if $r = 6$ in this formula, the surface area is 144π square meters.

38. D: Denote the width as *w* and the length as *l*. Then, $l = 3w + 5$. The perimeter is $2w + 2l = 90$. Substituting the first expression for *l* into the second equation yields $2(3w + 5) + 2w = 90$, or $8w = 80$, so w$l = 10$. Putting this into the first equation, yields $l = 3(10) + 5 = 35$.

39. C: To find the *y*-intercept, substitute zero for *x*, which gives:

$$y = 0^{\frac{5}{3}} + (0 - 3)(0 + 1)$$

$$0 + (-3)(1) = -3$$

40. E: The first field has an area of 1000 feet, so the length of one side is $\sqrt{1000} = 10\sqrt{10}$. Since there are four sides to a square, the total perimeter is $40\sqrt{10}$. The second square has an area of 10 square feet, so the length of one side is $\sqrt{10}$, and the total perimeter is $4\sqrt{10}$. Adding these together gives:

$$40\sqrt{10} + 4\sqrt{10}$$

$$(40 + 4)\sqrt{10} = 44\sqrt{10}$$

Mechanical Comprehension Test

The *Mechanical Comprehension (MC)* section tests a candidate's knowledge of mechanics and physical principles. These include concepts of force, energy, and work, and how they're used to predict the functioning of tools and machines. This knowledge is important for a successful career in the military. A good score on this section shows that a candidate has a solid background for learning how to use tools and machines properly. This is extremely important for the timely, safe completion of most tasks a future pilot or airman must undertake during his or her service.

The test problems in the MC section of the exam focus on understanding physical principles, but they are *qualitative* in nature rather than *quantitative*. This means the problems involve predicting the *behavior* of a system (such as the direction it moves) rather than calculating a specific measurement (such as its velocity). The figure below shows a sample problem similar to those on the MC test:

Mechanical Comprehension Sample Test Problem

Question 1.

Extending the reach of this crane will shift its

- **A.** total weight
- **B.** allowable speed
- **C.** center of gravity
- **D.** center of buoyancy

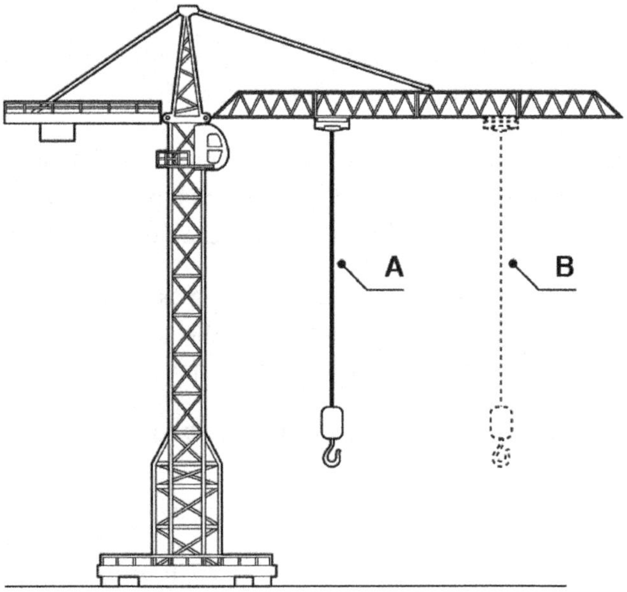

The sample problem pictures a system of a crane lifting a weight, and below the picture is a question. On the exam, it's *very important* to read these questions *carefully.* This question involves completing the following sentence: *Extending the reach of this crane will shift its* _____________. After the sentence, four possible answers are provided.

The correct answer is *C, center of gravity*. In this sample problem, it's easy to guess the correct answer simply by eliminating the rest. Answer *A* is incorrect because moving the load out along the crane's boom won't change its weight, just like moving a bodybuilder's arm that's holding a dumbbell won't change the combined weight of the bodybuilder and the dumbbell. Answer *B* is incorrect because the crane isn't moving. That leaves Answers *C* and *D*, but *D* is incorrect because buoyancy is only involved in systems with a liquid (the buoyancy of air is negligible). Therefore, through the process of elimination, *C* is the correct answer.

Review of Physics and Mechanical Principles

The proper use of tools and machinery depends on an understanding of basic physics, which includes the study of motion and the interactions of *mass*, *force*, and *energy*. These terms are used every day, but their exact meanings are difficult to define. In fact, they're usually defined in terms of each other.

The matter in the universe (atoms and molecules) is characterized in terms of its *mass*, which is measured in kilograms in the *International System of Units (SI)*. The amount of mass that occupies a given volume of space is termed *density*.

Mass occupies space, but it's also a component that inversely relates to acceleration when a force is applied to it. This *force* is the application of *energy* to an object with the intent of changing its position (mainly its acceleration).

To understand *acceleration*, it's necessary to relate it to displacement and velocity. The *displacement* of an object is simply the distance it travels. The *velocity* of an object is the distance it travels in a unit of time, such as miles per hour or meters per second:

$$Velocity = \frac{Distance\ Traveled}{Time\ Required}$$

There's often confusion between the words "speed" and "velocity." Velocity includes speed *and* direction. For example, a car traveling east and another traveling west can have the same speed of 30 miles per hour (mph), but their velocities are different. If movement eastward is considered positive, then movement westward is negative. Thus, the eastbound car has a velocity of 30 mph while the westbound car has a velocity of -30 mph.

The fact that velocity has a *magnitude* (speed) and a direction makes it a vector quantity. A *vector* is an arrow pointing in the direction of motion, with its length proportional to its magnitude.

Vectors can be added geometrically as shown below. In this example, a boat is traveling east at 4 *knots* (nautical miles per hour) and there's a current of 3 knots (thus a slow boat and a very fast current). If the boat travels in the same direction as the current, it gets a "lift" from the current and its speed is 7 knots. If the boat heads *into* the current, it has a forward speed of only 1 knot (4 knots – 3 knots = 1 knot) and makes very little headway. As shown in the figure below, the current is flowing north across the boat's path. Thus, for every 4 miles of progress the boat makes eastward, it drifts 3 miles to the north.

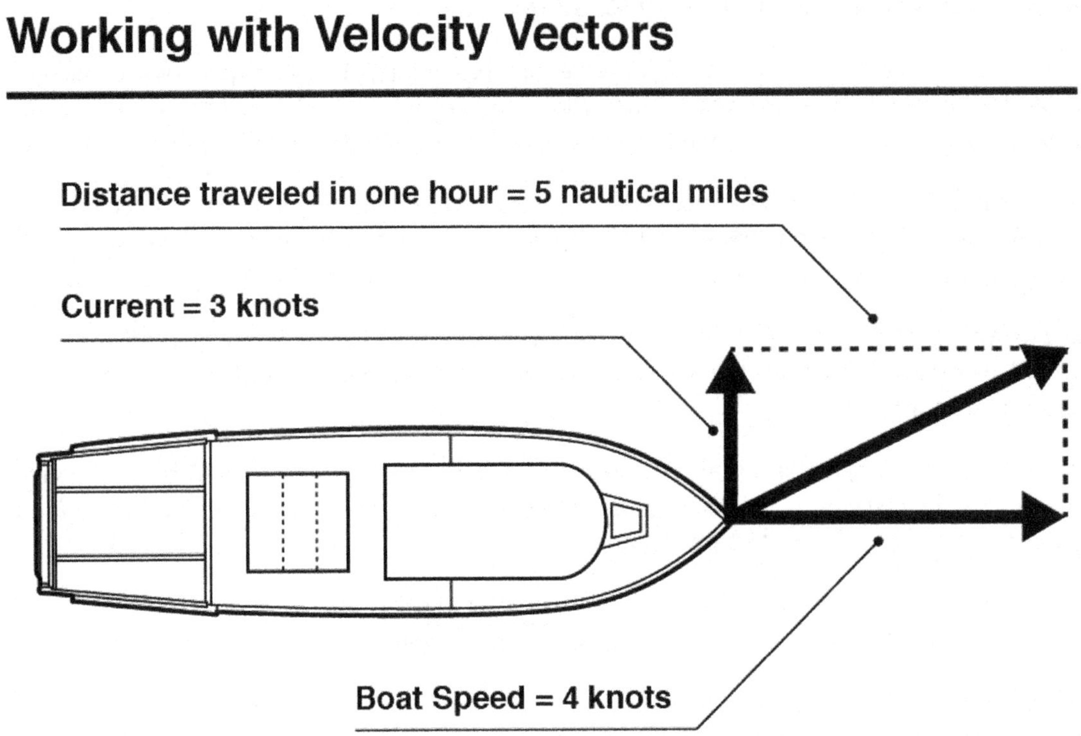

The total distance traveled is calculated using the *Pythagorean Theorem* for a right triangle, which should be memorized as follows:

$$a^2 + b^2 = c^2 \text{ or } c = \sqrt{a^2 + b^2}$$

Of course, the problem above was set up using a Pythagorean triple (3, 4, 5), which made the calculation easy.

Another example where velocity and speed are different is with a car traveling around a bend in the road. The speed is constant along the road, but the direction (and therefore the velocity) changes continuously.

The *acceleration* of an object is the change in its velocity in a given period of time:

$$Acceleration = \frac{Change\ in\ Velocity}{Time\ Required}$$

For example, a car starts at rest and then reaches a velocity of 70 mph in 8 seconds. What's the car's acceleration in feet per second squared?

First, the velocity must be converted from miles per hour to feet per second:

$$70\frac{miles}{hour} \times \frac{5{,}280\, feet}{mile} \times \frac{hour}{3600\ seconds} = 102.67\ feet/second$$

Starting from rest, the acceleration is:

$$Acceleration = \frac{102.67\frac{feet}{second} - 0\ \frac{feet}{second}}{8\ seconds} = 12.8\ feet/second^2$$

Newton's Laws

Isaac Newton's three laws of motion describe how the acceleration of an object is related to its mass and the forces acting on it. The three laws are:

1. Unless acted on by a force, a body at rest tends to remain at rest; a body in motion tends to remain in motion with a constant velocity and direction.
2. A force that acts on a body accelerates it in the direction of the force. The larger the force, the greater the acceleration; the larger the mass, the greater its inertia (resistance to movement and acceleration).
3. Every force acting on a body is resisted by an equal and opposite force.

To understand Newton's laws, it's necessary to understand forces. These forces can push or pull on a mass, and they have a magnitude and a direction. Forces are represented by a vector, which is the arrow lined up along the direction of the force with its tip at the point of application. The magnitude of the force is represented by the length of the vector.

The figure below shows a mass acted on or "pushed" by two equal forces (shown here by vectors of the same length). Both vectors "push" along the same line through the center of the mass, but in opposite directions. What happens?

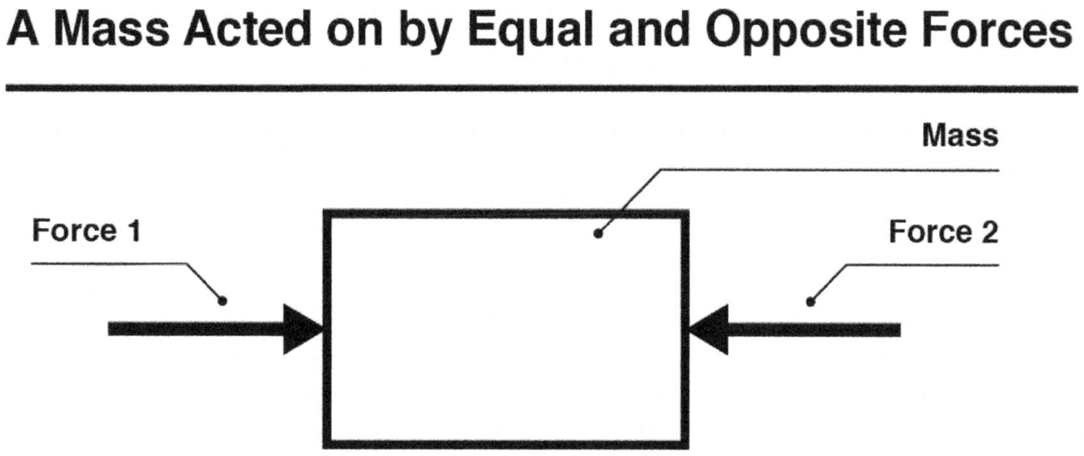

According to Newton's third law, every force on a body is resisted by an equal and opposite force. In the figure above, Force 1 acts on the left side of the mass. The mass pushes back. Force 2 acts on the right side, and the mass pushes back against this force too. The net force on the mass is zero, so according to Newton's first law, there's no change in the *momentum* (the mass times its velocity) of the mass. Therefore, if the mass is at rest before the forces are applied, it remains at rest. If the mass is in motion with a constant velocity, its momentum doesn't change. So, what happens when the net force on the mass isn't zero, as shown in the figure below?

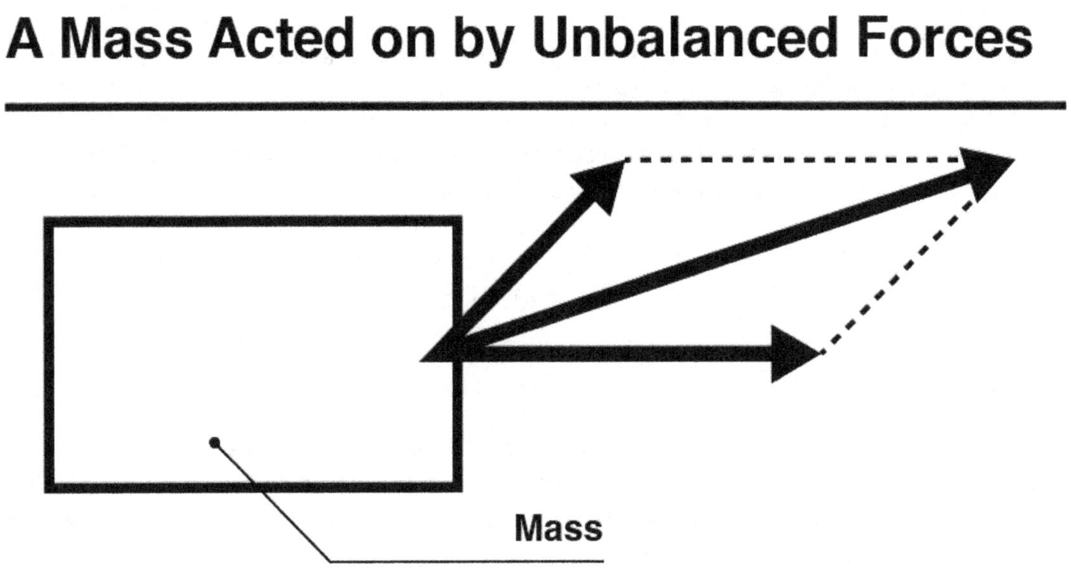

Notice that the forces are vector quantities and are added geometrically the same way that velocity vectors are manipulated.

Here in the figure above, the mass is pulled by two forces acting to the right, so the mass accelerates in the direction of the net force. This is described by Newton's second law:

Force = Mass x Acceleration

The force (measured in *newtons*) is equal to the product of the mass (measured in kilograms) and its acceleration (measured in meters per second squared or meters per second, per second). A better way to look at the equation is dividing through by the mass:

Acceleration = Force/Mass

This form of the equation makes it easier to see that the acceleration of an object varies directly with the net force applied and inversely with the mass. Thus, as the mass increases, the acceleration is reduced for a given force. To better understand, think of how a baseball accelerates when hit by a bat. Now imagine hitting a cannonball with the same bat and the same force. The cannonball is more massive than the baseball, so it won't accelerate very much when hit by the bat.

In addition to forces acting on a body by touching it, gravity acts as a force at a distance and causes all bodies in the universe to attract each other. The *force of gravity (F_g)* is proportional to the masses of the two objects (m and M) and inversely proportional to the square of the distance (r^2) between them (and G is the proportionality constant). This is shown in the following equation:

$$F_g = G\frac{mM}{r^2}$$

The force of gravity is what causes an object to fall to Earth when dropped from an airplane. Understanding gravity helps explain the difference between mass and weight. Mass is a property of an object that remains the same while it's intact, no matter where it's located. A 10-kilogram cannonball has the same mass on Earth as it does on the moon. On Earth, it *weighs* 98.1 newtons because of the attractive force of gravity, so it accelerates at 9.81 m/s^2. However, on the moon, the same cannonball has a weight of only about 16 newtons. This is because the gravitational attraction on the moon is approximately one-sixth that on Earth. Although Earth still attracts the body on the moon, it's so far away that its force is negligible.

For Americans, there's often confusion when talking about mass because the United States still uses "pounds" as a measurement of weight. In the traditional system used in the United States, the unit of mass is called a *slug*. It's derived by dividing the weight in pounds by the acceleration of gravity (32 feet/s^2); however, it's rarely used today. To avoid future problems, test takers should continue using SI units and *remember to express mass in kilograms and weight in Newtons*.

Another way to understand Newton's second law is to think of it as an object's change in momentum, which is defined as the product of the object's mass and its velocity:

Momentum = Mass x Velocity

Which of the following has the greater momentum: a pitched baseball, a softball, or a bullet fired from a rifle?

A bullet with a mass of 5 grams (0.005 kilograms) is fired from a rifle with a muzzle velocity of 2200 mph. Its momentum is calculated as:

$$2200\frac{miles}{hour}\times\frac{5{,}280\,feet}{mile}\times\frac{m}{3.28\,feet}\times\frac{hour}{3600\,seconds}\times 0.005kg = 4.92\frac{kg.m}{seconds}$$

A softball has a mass between 177 grams and 198 grams and is thrown by a college pitcher at 50 miles per hour. Taking an average mass of 188 grams (0.188 kilograms), a softball's momentum is calculated as:

$$50\frac{miles}{hour}\times\frac{5280\,feet}{mile}\times\frac{m}{3.28\,ft}\times\frac{hour}{3600\,seconds}\times 0.188kg = 4.19\frac{kg.m}{seconds}$$

That's only slightly less than the momentum of the bullet. Although the speed of the softball is considerably less, its mass is much greater than the bullet's.

A professional baseball pitcher can throw a 145-gram baseball at 100 miles per hour. A similar calculation (try doing it!) shows that the pitched hardball has a momentum of about 6.48 kg.m/seconds. That's more momentum than a speeding bullet!

So why is the bullet more harmful than the hard ball? It's because the force that it applies acts on a much smaller area.

Instead of using acceleration, Newton's second law is expressed here as the change in momentum (with the delta symbol "Δ" meaning "change"):

$$Force = \frac{\Delta\, Momentum}{\Delta\, Time} = \frac{\Delta\,(Mass \times Velocity)}{\Delta\, Time} = Mass \times \frac{\Delta\, Velocity}{\Delta\, Time}$$

The rapid application of force is called *impulse*. Another way of stating Newton's second law is in terms of the impulse, which is the force multiplied by its time of application:

$$Impluse = Force \times \Delta\, Time = Mass \times \Delta\, Velocity$$

In the case of the rifle, the force created by the pressure of the charge's explosion in its shell pushes the bullet, accelerating it until it leaves the barrel of the gun with its *muzzle velocity* (the speed the bullet has when it leaves the muzzle). After leaving the gun, the bullet doesn't accelerate because the gas pressure is exhausted. The bullet travels with a constant velocity in the direction it's fired (ignoring the force exerted against the bullet by friction and drag).

Similarly, the pitcher applies a force to the ball by using their muscles when throwing. Once the ball leaves the pitcher's fingers, it doesn't accelerate and the ball travels toward the batter at a constant speed (again ignoring friction and drag). The speed is constant, but the velocity can change if the ball travels along a curve.

Projectile Motion

According to Newton's first law, if no additional forces act on the bullet or ball, it travels in a straight line. This is also true if the bullet is fired in outer space. However, here on Earth, the force of gravity continues to act so the motion of the bullet or ball is affected.

What happens when a bullet is fired from the top of a hill using a rifle held perfectly horizontal? Ignoring air resistance, its horizontal velocity remains constant at its muzzle velocity. Its vertical velocity (which is zero when it leaves the gun barrel) increases because of gravity's acceleration. Each passing second, the bullet traces out the same distance horizontally while increasing distance vertically (shown in the figure below). In the end, the projectile traces out a *parabolic curve*.

Projectile Path for a Bullet Fired Horizontally from a Hill (Ignoring Air Resistance)

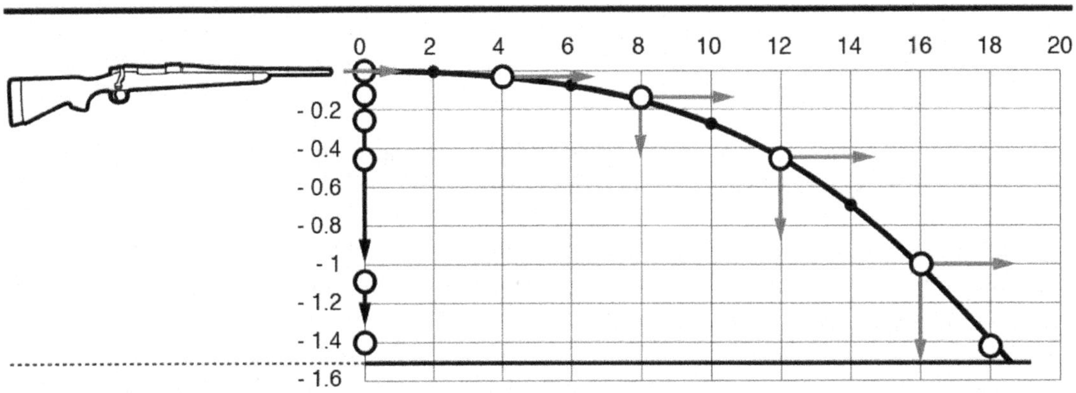

This vertical, downward acceleration is why a pitcher must put an arc on the ball when throwing across home plate. Otherwise the ball will fall at the batter's feet.

It's also interesting to note that if an artillery crew simultaneously drops one cannonball and fires another one horizontally, the two cannonballs will hit the ground at the same time since both balls are accelerating at the same rate and experience the same changes in vertical velocity.

What if air resistance is taken into account? This is best answered by looking at the horizontal and vertical motions separately.

The horizontal velocity is no longer constant because the initial velocity of the projectile is continually reduced by the resistance of the air. This is a complex problem in fluid mechanics, but it's sufficient to note that that the projectile doesn't fly as far before landing as predicted from the simple theory.

The vertical velocity is also reduced by air resistance. However, unlike the horizontal motion where the propelling force is zero after the cannonball is fired, the downward force of gravity acts continuously. The downward velocity increases every second due to the acceleration of gravity. As the velocity increases, the resisting force (called *drag*) increases with the square of the velocity. If the projectile is fired or dropped from a sufficient height, it reaches a terminal velocity such that the upward drag force equals the downward force of gravity. When that occurs, the projectile falls at a constant rate.

This is the same principle that's used for a parachute. Its drag (caused by its shape that scoops up air) is sufficient enough to slow down the fall of the parachutist to a safe velocity, thus avoiding a fatal crash on the ground.

So, what's the bottom line? If the vertical height isn't too great, a real projectile will fall short of the theoretical point of impact. However, if the height of the fall is significant and the drag of the object results in a small terminal fall velocity, then the projectile can go further than the theoretical point of impact.

What if the projectile is launched from a moving platform? In this case, the platform's velocity is added to the projectile's velocity. That's why an object dropped from the mast of a moving ship lands at the base of the mast rather than behind it. However, to an observer on the shore, the object traces out a parabolic arc.

Angular Momentum

In the previous examples, all forces acted through the center of the mass, but what happens if the forces aren't applied through the same line of action, like in the figure below?

A Mass Acted on by Forces Out of Line with Each Other

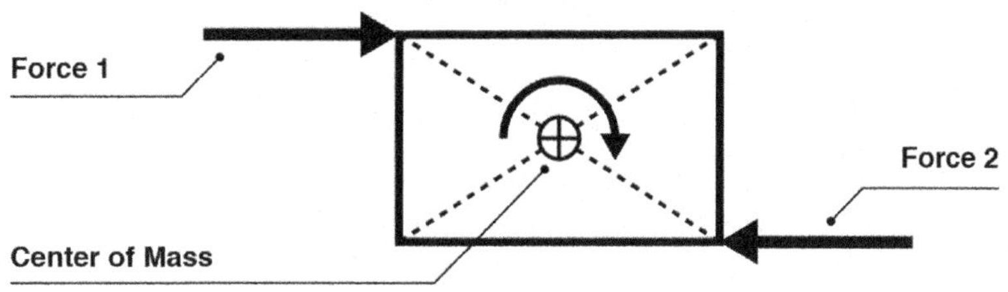

When this happens, the two forces create *torque* and the mass rotates around its center of gravity. In the figure above, the center of gravity is the center of the rectangle ("Center of Mass"), which is determined by the two, intersecting main diagonals. The center of an irregularly shaped object is found by hanging it from two different edges, and the center of gravity is at the intersection of the two "plumb lines."

Newton's second law still applies when the forces form a moment pair, but it must be expressed in terms of angular acceleration and the moment of inertia. The *moment of inertia* is a measure of the body's resistance to rotation, similar to the mass's resistance to linear acceleration. The more compact the body, the less the moment of inertia and the faster it rotates, much like how an ice skater spinning with outstretched arms will speed up as the arms are brought in close to the body.

The concept of torque is important in understanding the use of wrenches and is likely to be on the test. The concept of torque and moment/lever arm will be taken up again below, when the physics of simple machines is presented.

Energy and Work

The previous examples of moving boats, cars, bullets, and baseballs are examples of simple systems that are thought of as particles with forces acting through their center of gravity. They all have one property in common: *energy*. The energy of the system results from the forces acting on it and is considered its ability to do work.

Work or the energy required to do work (which are the same) is calculated as the product of force and distance traveled along the line of action of the force. It's measured in *foot-pounds* in the traditional system (which is still used in workshops and factories) and in *newton meters (N·m)* in the International System of Units (SI), which is the preferred system of measurement today.

Potential and Kinetic Energy

Energy can neither be created nor destroyed, but it can be converted from one form to another. There are many forms of energy, but it's useful to start with mechanical energy and potential energy.

The *potential energy* of an object is equal to the work that's required to lift it from its original elevation to its current elevation. This is calculated as the weight of the object or its downward force (mass times the acceleration of gravity) multiplied by the distance (y) it is lifted above the reference elevation or "datum." This is written:

$$PE = mgy$$

The mechanical or *kinetic energy* of a system is related to its mass and velocity:

$$KE = \frac{1}{2}mv^2$$

The *total energy* is the sum of the kinetic energy and the potential energy, both of which are measured in foot-pounds or newton meters.

If a weight with a mass of 10 kilograms is raised up a ladder to a height of 10 meters, it has a potential energy of $10\text{m} \times 10\text{kg} \times 9.81\text{m/s}^2 = 981\text{N}\cdot\text{m}$. This is approximately 1000 newton meters if the acceleration of gravity (9.81 m/s^2) is rounded to 10 m/s^2, which is accurate enough for most earth-bound calculations. It has zero kinetic energy because it's at rest, with zero velocity.

If the weight is dropped from its perch, it accelerates downward so that its velocity and kinetic energy increase as its potential energy is "used up" or, more precisely, converted to kinetic energy.

When the weight reaches the bottom of the ladder, just before it hits the ground, it has a kinetic energy of 981 N·m (ignoring small losses due to air resistance). The velocity can be solved by using the following:

$$981\ N \cdot m = \frac{1}{2} 10\ kg\ \times\ v^2 \quad \text{or} \quad v = 14.01\ m/s$$

When the 10-kilogram weight hits the ground, its potential energy (which was measured *from* the ground) and its velocity are both zero, so its kinetic energy is also zero. What's happened to the energy? It's dissipated into heat, noise, and kicking up some dust. It's important to remember that energy can neither be created nor destroyed, so it can only change from one form to another.

The conversion between potential and kinetic energy works the same way for a pendulum. If it's raised and held at its highest position, it has maximum potential energy but zero kinetic energy.

Potential and Kinetic Energy for a Swinging Pendulum

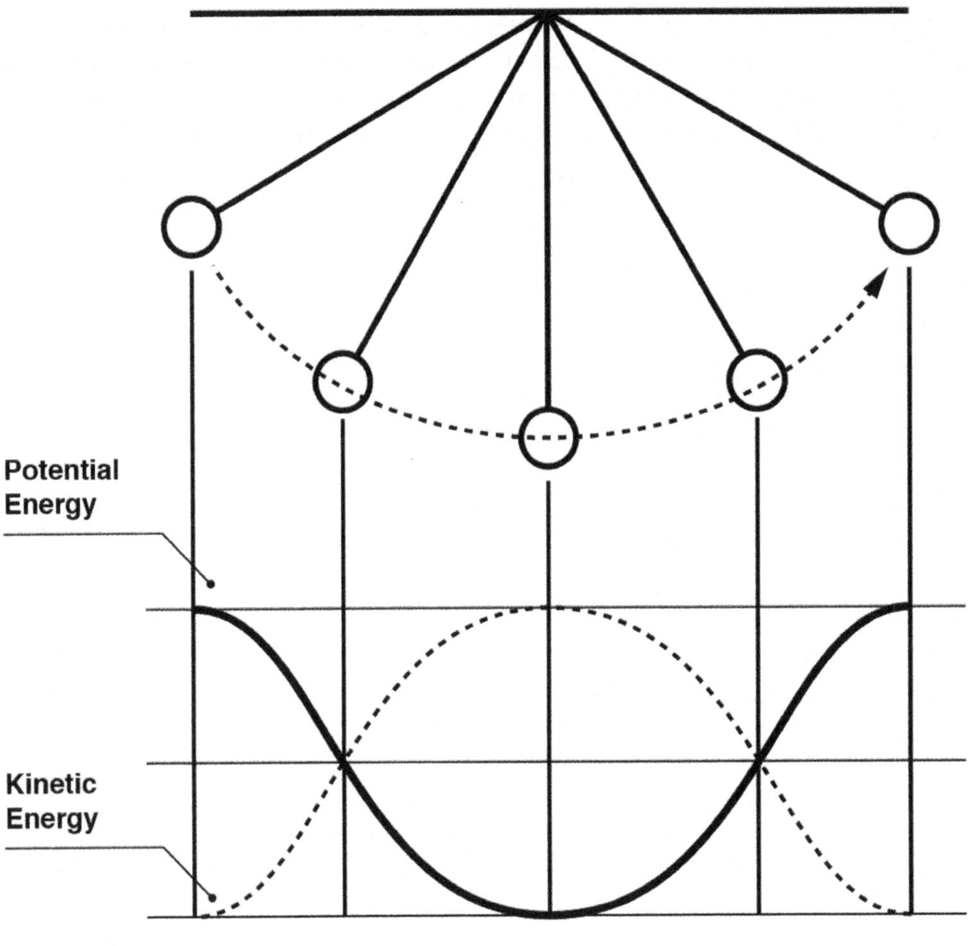

When the pendulum is released from its highest position (see left side of the figure above), it swings down so that its kinetic energy increases as its potential energy decreases. At the bottom of its swing, the pendulum is moving at its maximum velocity with its maximum kinetic energy. As the pendulum swings past the bottom of its path, its velocity slows down as its potential energy increases.

Work

The released potential energy of a system can be used to do *work*.

For instance, most of the energy lost by letting a weight fall freely can be recovered by hooking it up to a pulley to do work by pulling another weight back up (as shown in the figure below).

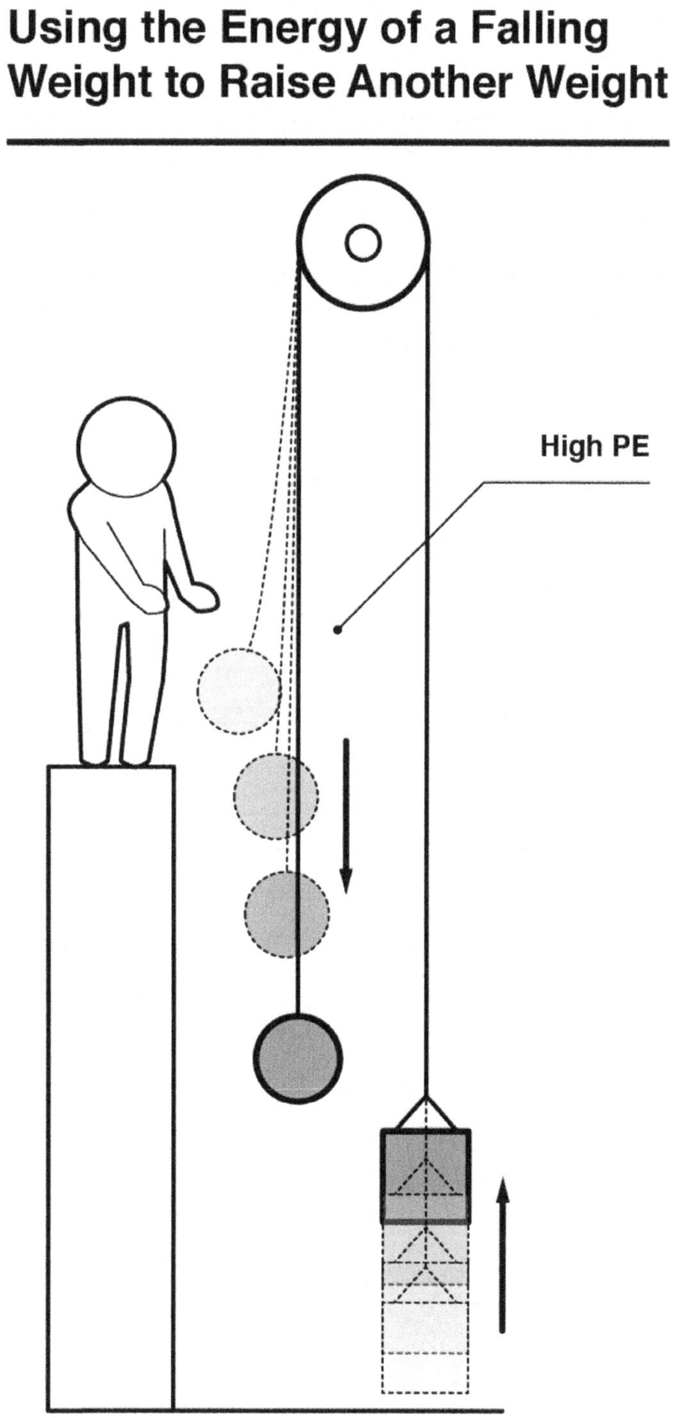

In other words, the potential energy expended to lower the weight is used to do the work of lifting another object. Of course, in a real system, there are losses due to friction. The action of pulleys will be discussed later in this study guide.

Since *energy* is defined as *the capacity to do work*, energy and work are measured in the same units:

$$Energy = Work = Force \times Distance$$

Force is measured in *newtons (N)*. Distance is measured in meters. The units of work are *newton meters (N·m)*. The same is true for kinetic energy and potential energy.

Another way to store energy is to compress a spring. Energy is stored in the spring by stretching or compressing it. The work required to shorten or lengthen the spring is given by the equation:

$$F = k \times d$$

Here, "d" is the length in meters and "k" is the resistance of the spring constant (measured in N·m), which is a constant as long as the spring isn't stretched past its elastic limit. The resistance of the spring is constant, but the force needed to compress the spring increases with each millimeter it's pushed.

The potential energy stored in the spring is equal to the work done to compress it, which is the total force times the change in length. Since the resisting force of the spring increases as its displacement increases, the average force must be used in the calculation:

$$W = PE = F \times d$$

$$\frac{1}{2}(F_i + F_f)d \times d$$

$$\frac{1}{2}(0 + F_f)d \times d = \frac{1}{2}Fd^2$$

The potential energy in the spring is stored by locking it into place, and the work energy used to compress it is recovered when the spring is unlocked. It's the same when dropping a weight from a height—the energy doesn't have to be wasted. In the case of the spring, the energy is used to propel an object.

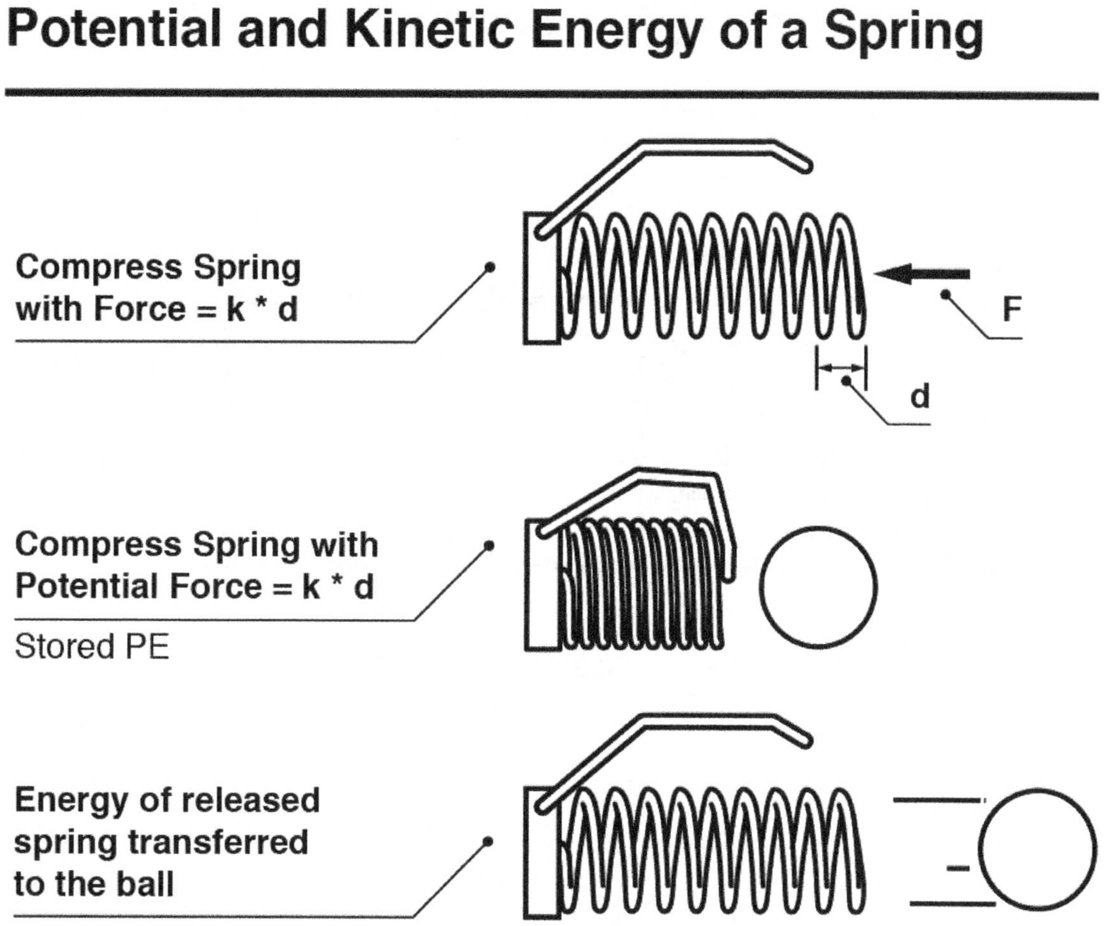

Pushing a block horizontally along a rough surface requires work. In this example, the work needs to overcome the force of friction, which opposes the direction of the motion and equals the weight of the block times a *friction factor (f)*. The friction factor is greater for rough surfaces than smooth surfaces, and it's usually greater *before* the motion starts than after it has begun to slide. These terms are illustrated in the figure below.

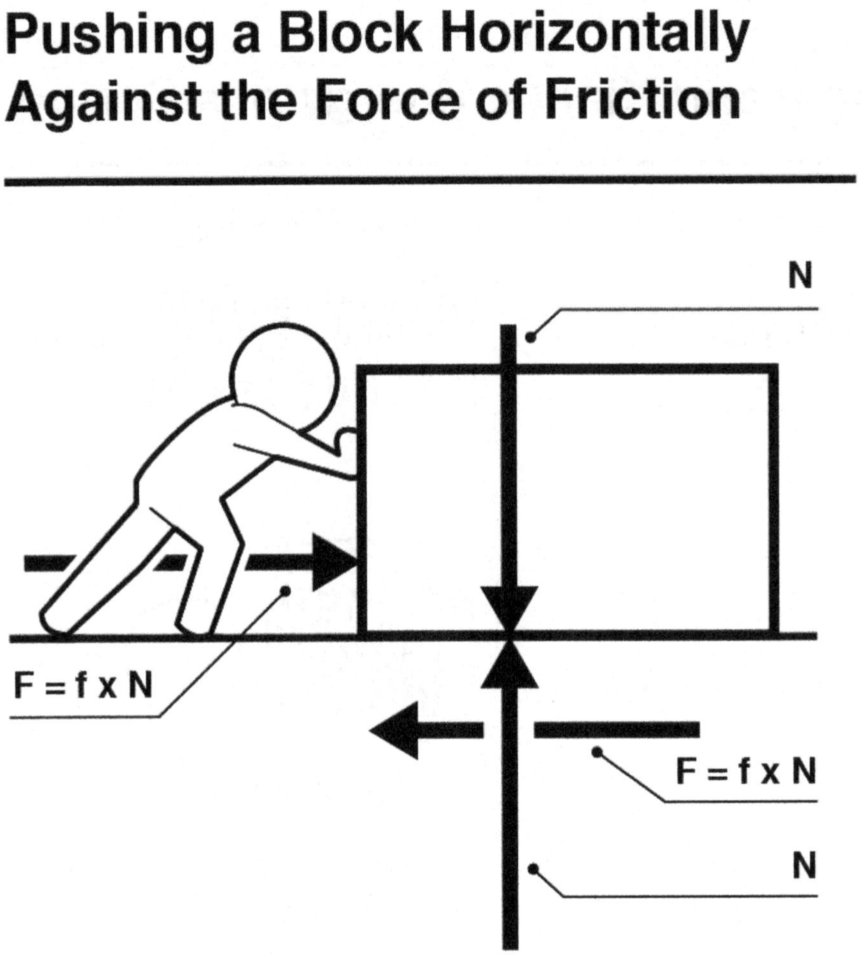

When pushing a block, there's no increase in potential energy since the block's elevation doesn't change. Expending the energy to overcome friction is "wasted" in the generation of heat. Yet, to move a block from point A to point B, an energy cost must be paid. However, friction isn't always a hindrance. In fact, it's the force that makes the motion of a wheel possible.

Heat energy can also be created by burning organic fuels, such as wood, coal, natural gas, and petroleum. All of these are derived from plant matter that's created using solar energy and photosynthesis. The chemical energy or *"heat"* liberated by the combustion of these fuels is used to warm buildings during the winter or even melt metal in a foundry. The heat is also used to generate steam, which can drive engines or turn turbines to generate electric energy.

In fact, work and heat are interchangeable. This fact was first recognized by gun founders when they were boring out cast, brass cannon blanks. The cannon blanks were submerged in a water bath to reduce friction, yet as the boring continued, the water bath boiled away!

Later, the amount of work needed to raise the temperature of water was measured by an English physicist (and brewer) named James Prescott Joule. The way that Joule measured the mechanical equivalent of heat is illustrated in the figure below. This setup is similar to the one in the figure above with the pulley, except instead of lifting another weight, the falling weight's potential energy is converted to the mechanical energy of the rotating vertical shaft. This turns the paddles, which churns the water to increase its temperature. Through a long series of repeated measurements, Joule showed that 4186 N·m of work was necessary to raise the temperature of one kilogram of water by one degree Celsius, no matter how the work was delivered.

Device Measuring the Mechanical Energy Needed to Increase the Temperature of Water

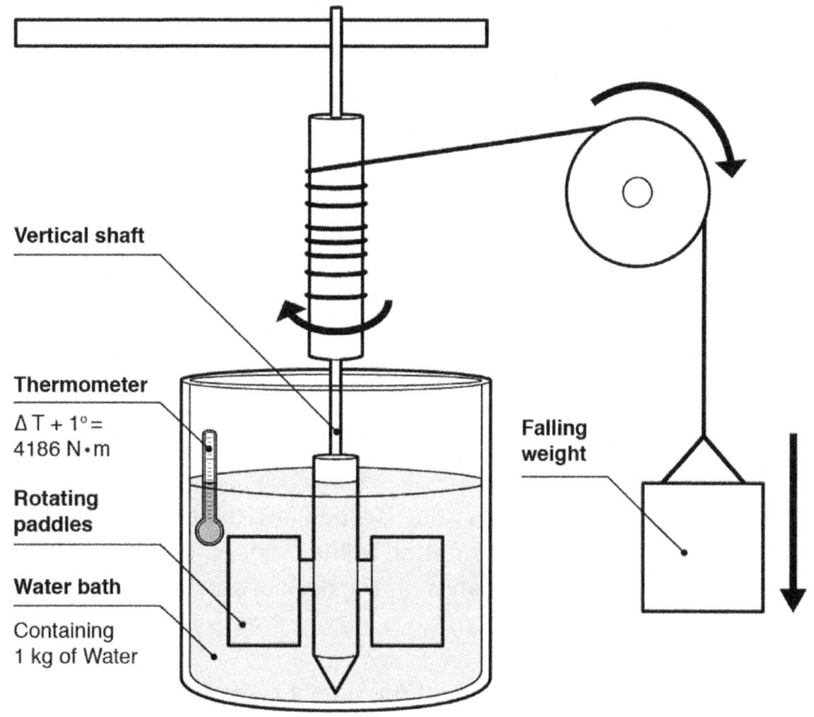

In recognition of this experiment, the newton meter is also called a "*joule*." Linking the names for work and heat to the names of two great physicists is truly appropriate because heat and work being interchangeable is of the greatest practical importance. These two men were part of a very small, select group of scientists for whom units of measurement have been named: Marie Curie for radioactivity, Blaise Pascal for pressure, James Watt for power, Andre Ampere for electric current, and only a few others.

Just as mechanical work is converted into heat energy, heat energy is converted into mechanical energy in the reverse process. An example of this is a closely fitting piston supporting a weight and mounted in a cylinder where steam enters from the bottom.

In this example, water is heated into steam in a boiler, and then the steam is drawn off and piped into a cylinder. Steam pressure builds up in the piston, exerting a force in all directions. This is counteracted by the tensile strength of the cylinder; otherwise, it would burst. The pressure also acts on the exposed face of the piston, pushing it upwards against the load (displacing it) and thus doing work.

Work developed from the pressure acting over the area exerts a force on the piston as described in the following equation:

$$Work = Pressure \times Piston\ Area \times Displacement$$

Here, the work is measured in newton meters, the pressure in newtons per square meter or *pascals (Pa)*, and the piston displacement is measured in meters.

Since the volume enclosed between the cylinder and piston increases with the displacement, the work can also be expressed as:

$$Work = Pressure \times \Delta\,Volume$$

For example, a 10-kilogram weight is set on top of a piston-cylinder assembly with a diameter of 25 centimeters. The area of the cylinder is:

$$Area = \frac{\pi \times d^2}{4} = 0.785 \times 0.25^2 = .049\,m^2$$

If the acceleration due to gravity is approximately 10 m/s^2, and the area is rounded to .05 meters squared, then the pressure needed to counteract the weight of the 10-kilogram weight is estimated as:

$$P = \frac{F}{A} \approx 10 \times \frac{10}{0.05} = 2000\frac{N}{m^2} = 2000\,Pa = 2\,KPa$$

If steam with a pressure slightly greater than this value is piped into the cylinder, it slowly lifts the load. If steam at a much higher pressure is suddenly admitted to the cylinder, it throws the load into the air. This is the principle used to steam-catapult airplanes off the deck of an aircraft carrier.

Power

Power is defined as the rate at which work is done, or the time it takes to do a given amount of work. In the International System of Units (SI), work is measured in *newton meters (N·m)* or *joules (J)*. Power is measured in joules/second or *watts (W)*.

For example, to raise a 1-kilogram mass one meter off the ground, it takes approximately 10 newton meters of work (approximating the gravitational acceleration of 9.81 m/s^2 as 10 m/s^2). To do the work in 10 seconds, it requires 1 watt of power. Doing it in 1 second requires 10 watts of power. Essentially, *doing it faster means dividing by a smaller number*, and that means greater power.

Although SI units are preferred for technical work throughout the world, the old traditional (or English) unit of measuring power is still used. Introduced by *James Watt* (the same man for whom the SI unit of power "watt" is named), the unit of *horsepower (HP)* rated the power of the steam engines that he and his partner (Matthew Boulton) manufactured and sold to mine operators in 18th century England. The mine operators used these engines to pump water out of flooded facilities in the beginning of the Industrial Revolution.

To provide a measurement that the miners would be familiar with, Watt and Boulton referenced the power of their engines with the "power of a horse."

Watt's measurements showed that, on average, a well-harnessed horse could lift a 330-pound weight 100 feet up a well in one minute (330 pounds is the weight of a 40-gallon barrel filled to the brim). Remembering that power is expressed in terms of energy or work per unit time, horsepower came to be measured as:

$$1\ HP = \frac{100\ feet\ \times\ 330\ pounds}{1\ minute} \times \frac{1\ minute}{60\ seconds} = 550\ foot\ pounds/second$$

A horse that pulled the weight up faster, or pulled up more weight in the same time, was a more *powerful* horse than Watt's "average horse."

Hundreds of millions of engines of all types have been built since Watt and Boulton started manufacturing their products, and the unit of *horsepower* has been used throughout the world to this day. Of course, modern technicians and engineers still need to convert horsepower to watts to work with SI units. An approximate conversion is *1 HP = 746 W.*

Take for example a 2016 CTS-V Cadillac rated at 640 HP. If a *megawatt (MW)* is one million watts, that means the Cadillac has almost half a megawatt of power as shown by this conversion:

$$640\ HP \times \frac{746\ W}{1\ HP} = 477{,}440\ W = 477.4\ kW$$

The power of the Cadillac is comparable to that of the new Westinghouse AP-1000 Nuclear Power Plant, which is rated at 1100 MW or the equivalent of 2304 Cadillacs (assuming no loss in power). That would need a very big parking lot and a tremendous amount of gasoline!

A question that's often asked is, "How much energy is expended by running an engine for a fixed amount of time?" This is important to know when planning how much fuel is needed to run an engine. For instance, how much energy is expended in running the new Cadillac at maximum power for 30 minutes?

In this case, the energy expenditure is approximately 240 kilowatt hours. This must be converted to joules, using the conversion factor that one watt equals one joule per second:

$$240{,}000\ W\ hours \times \frac{3600\ seconds}{1\ hour} = 8.64(10)^8\ joules$$

So how much gasoline is burned? Industrial tests show that a gallon of gasoline is rated to contain about 1.3 x 10^8 joules of energy. That's 130 million joules per gallon. The gallons of gasoline are obtained by dividing:

$$\frac{8.64(10)^8 J}{1.3(10)^8 J/gallon} = 6.65\ gallons \times \frac{3.8\ liters}{gallon} = 25.3\ liters$$

The calculation has now come full circle. It began with power. Power equals energy divided by time. Power multiplied by time equals the energy needed to run the machine, which came from burning fuel.

Fluids

In addition to the behavior of solid particles acted on by forces, it is important to understand the behavior of fluids. Fluids include both liquids and gasses. The best way to understand fluid behavior is to contrast it with the behavior of solids, as shown in the figure below.

First, consider a block of ice, which is solid water. If it is set down inside a large box it will exert a force on the bottom of the box due to its weight as shown on the left, in Part A of the figure. The solid block exerts a pressure on the bottom of the box equal to its total weight divided by the area of its base:

$$Pressure = Weight\ of\ block/Area\ of\ base$$

That pressure acts only in the area directly under the block of ice.

If the same mass of ice is melted, it behaves much differently. It still has the same weight as before because its mass hasn't changed. However, the volume has decreased because liquid water molecules are more tightly packed together than ice molecules, which is why ice floats (it is less dense).

The Behavior of Solids and Liquids Compared

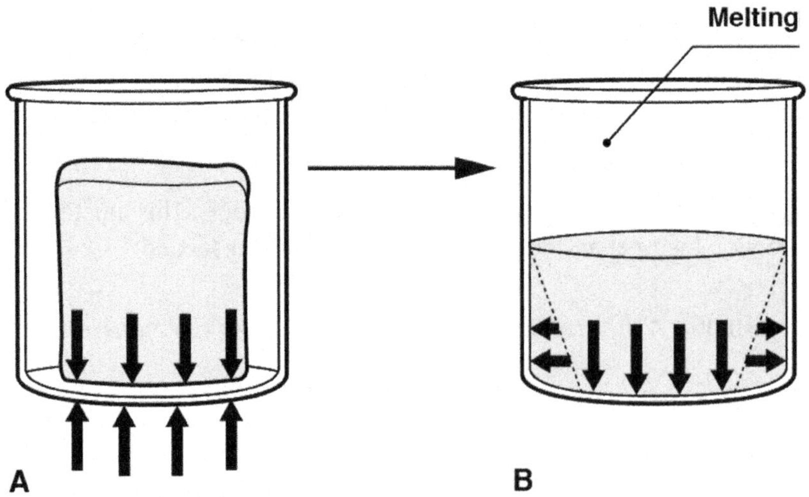

A — Forces exerted by a block of ice on its container

B — Forces exerted by a water on its container

The melted ice (now water) conforms to the shape of the container. This means that the fluid exerts pressure not only on the base, but on the sides of the box at the water line and below. Actually, pressure in a liquid is exerted in all directions, but all the forces in the interior of the fluid cancel each other out, so that a net force is only exerted on the walls. Note also that the pressure on the walls increases with the depth of the water.

The fact that the liquid exerts pressure in all directions is part of the reason some solids float in liquids. Consider the forces acting on a block of wood floating in water, as shown in the figure below.

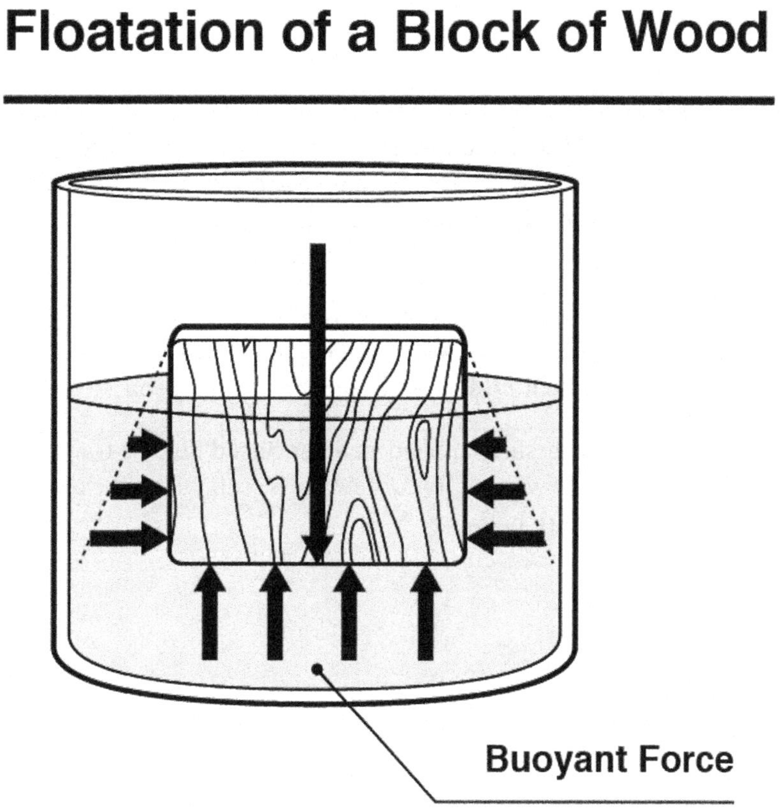

The block of wood is submerged in the water and pressure acts on its bottom and sides as shown. The weight of the block tends to force it down into the water. The force of the pressure on the left side of the block just cancels the force of the pressure on the right side.

There is a net upward force on the bottom of the block due to the pressure of the water acting on that surface. This force, which counteracts the weight of the block, is known as the *buoyant force*.

The block will sink to a depth such that the buoyant force of the water (equal to the weight of the volume displaced) just matches the total weight of the block. This will happen if two conditions are met:

1. The body of water is deep enough to float the block
2. The density of the block is less than the density of the water

If the body of water is not deep enough, the water pressure on the bottom side of the block won't be enough to develop a buoyant force equal to the block's weight. The block will be "beached" just like a boat caught at low tide.

If the density of the block is greater than the density of the fluid, the buoyant force acting on the bottom of the boat will not be sufficient to counteract the total weight of the block. That's why a solid steel block will sink in water.

If steel is denser than water, how can a steel ship float? The steel ship floats because it's hollow. The volume of water displaced by its steel shell (hull) is heavier than the entire weight of the ship and its contents (which includes a lot of empty space). In fact, there's so much empty space within a steel ship's hull that it can bob out of the water and be unstable at sea if some of the void spaces (called ballast tanks) aren't filled with water. This provides more weight and balance (or "trim") to the vessel.

The discussion of buoyant forces on solids holds for liquids as well. A less dense liquid can float on a denser liquid if they're *immiscible* (do not mix). For instance, oil can float on water because oil isn't as dense as the water. Fresh water can float on salt water for the same reason.

Pascal's law states that a change in pressure, applied to an enclosed fluid, is transmitted undiminished to every portion of the fluid and to the walls of its containing vessel. This principle is used in the design of hydraulic jacks, as shown in the figure below.

A force (F_1) is exerted on a small "driving" piston, which creates pressure on the hydraulic fluid. This pressure is transmitted through the fluid to a large cylinder. While the pressure is the same everywhere in the oil, the pressure action on the area of the larger cylinder creates a much higher upward force (F_2).

Illustration of a Hydraulic Jack Exemplifying Pascal's Law

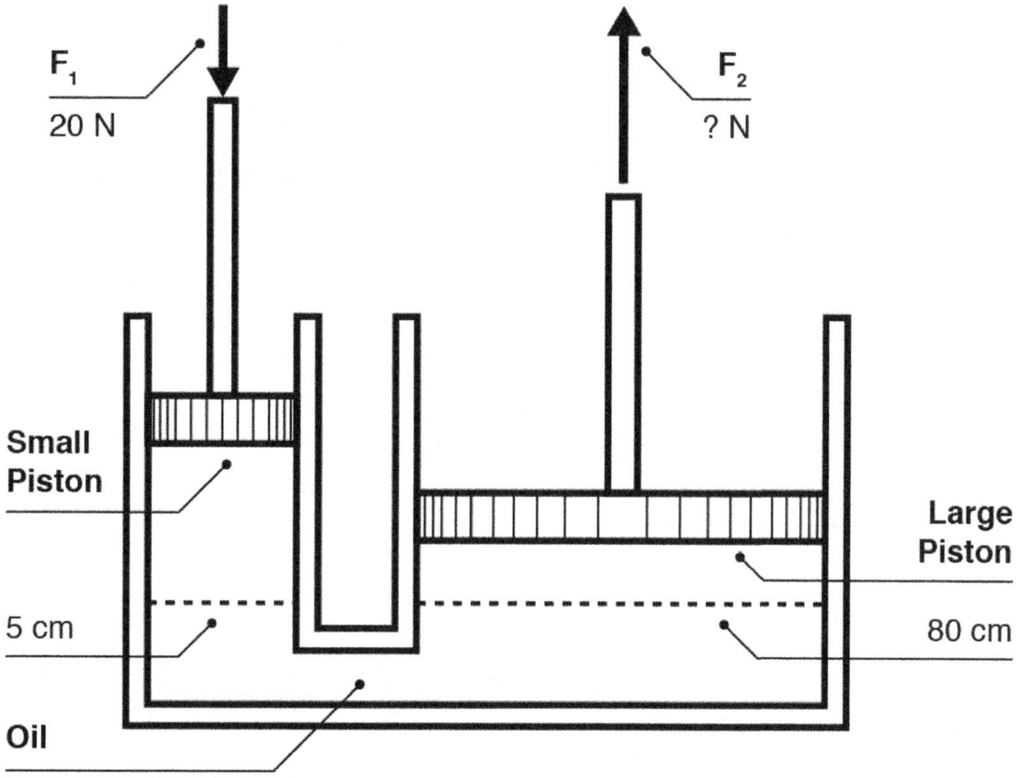

Looking again at the figure above, suppose the diameter of the small cylinder is 5 centimeters and the diameter of the large cylinder is 80 centimeters. If a force of 20 newtons (N) is exerted on the small driving piston, what's the value of the upward force F_2? In other words, what weight can the large piston support?

The pressure within the system is created from the force F_1 acting over the area of the piston:

$$P = \frac{F_1}{A} = \frac{20\ N}{\pi\,(0.05\ m)^2/4} = 10{,}185\ Pa$$

The same pressure acts on the larger piston, creating the upward force, F_2:

$$F_2 = P \times A = 10{,}185\ Pa \times \pi \times (0.8\ m)^2/4 = 5120\ N$$

Because a liquid has no internal shear strength, it can be transported in a pipe or channel between two locations. A fluid's "rate of flow" is the volume of fluid that passes a given location in a given amount of time and is expressed in *m^3/second*. The *flow rate* (*Q*) is determined by measuring the *area of flow* (*A*) in *m^2*, and the *flow velocity* (*v*) in *m/s*:

$$Q = v \times A$$

This equation is called the *Continuity Equation*. It's one of the most important equations in engineering and should be memorized. For example, what is the flow rate for a pipe with an inside diameter of 1200 millimeters running full with a velocity of 1.6 m/s (measured by a *sonic velocity meter*)?

Using the Continuity Equation, the flow is obtained by keeping careful track of units:

$$Q = v \times A = 1.6\frac{m}{s} \times \frac{\pi}{4} \times \left(\frac{1200\ mm}{1000\ mm/m}\right)^2 = 1.81\ m^3/second$$

For more practice, imagine that a pipe is filling a storage tank with a diameter of 100 meters. How long does it take for the water level to rise by 2 meters?

Since the flow rate (*Q*) is expressed in m^3/second, and volume is measured in m^3, then the time in seconds to supply a volume (*V*) is *V/Q*. Here, the volume required is:

$$Volume\ Required = Base\ Area \times Depth$$

$$\frac{\pi}{4}100^2 \times 2\ m = 15{,}700\ m^3$$

Thus, the time to fill the tank another 2 meters is 15,700 *m^3* divided by 1.81 *m^3/s* = 8674 seconds or 2.4 hours.

It's important to understand that, for a given flow rate, a smaller pipe requires a higher velocity.

The energy of a flow system is evaluated in terms of potential and kinetic energy, the same way the energy of a falling weight is evaluated. The total energy of a fluid flow system is divided into potential energy of elevation, and pressure and the kinetic energy of velocity. *Bernoulli's Equation* states that, for a constant flow rate, the total energy of the system (divided into components of elevation, pressure, and velocity) remains constant. This is written as:

$$Z + \frac{P}{\rho g} + \frac{v^2}{2g} = Constant$$

Each of the terms in this equation has dimensions of meters. The first term is the *elevation energy*, where *Z* is the elevation in meters. The second term is the *pressure energy*, where *P* is the pressure, *ρ* is the density, and *g* is the acceleration of gravity. The dimensions of the second term are also in meters. The third term is the *velocity energy*, also expressed in meters.

For a fixed elevation, the equation shows that, as the pressure increases, the velocity decreases. In the other case, as the velocity increases, the pressure decreases.

The use of the Bernoulli Equation is illustrated in the figure below. The total energy is the same at Sections 1 and 2. The area of flow at Section 1 is greater than the area at Section 2. Since the flow rate is the same at each section, the velocity at Point 2 is higher than at Point 1:

$$Q = V_1 \times A_1 = V_2 \times A_2, \qquad V_2 = V_1 \times \frac{A_1}{A_2}$$

Finally, since the total energy is the same at the two sections, the pressure at Point 2 is less than at Point 1. The tubes drawn at Points 1 and 2 would actually have the water levels shown in the figure; the pressure at each point would support a column of water of a height equal to the pressure divided by the unit weight of the water ($h = P/\rho g$).

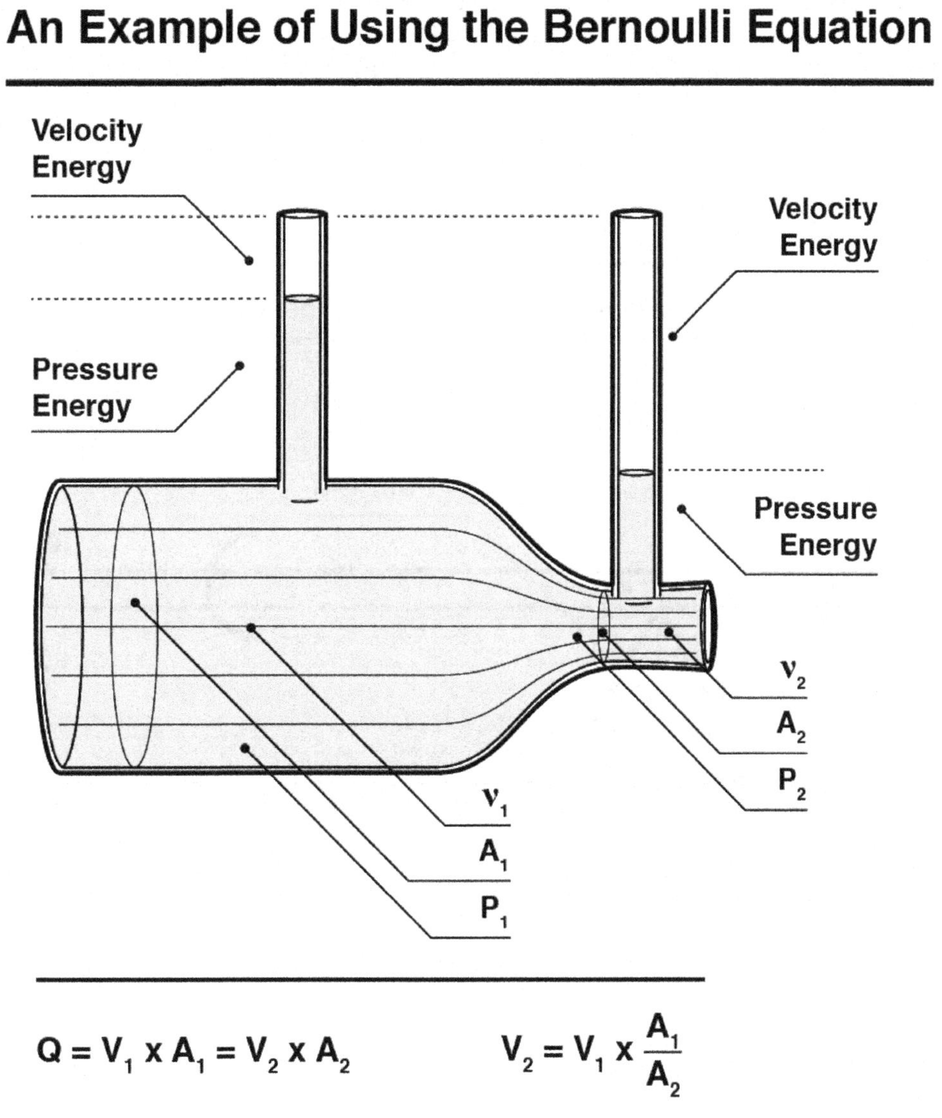

Machines

Now that the basic physics of work and energy have been discussed, the common machines used to do the work can be discussed in more detail.

A *machine* is a device that: transforms energy from one form to another, multiplies the force applied to do work, changes the direction of the resultant force, or increases the speed at which the work is done.

The details of how energy is converted into work by a system are extremely complicated but, no matter how complicated the "linkage" between the components, every system is composed of certain elemental or simple machines. These are discussed briefly in the following sections.

Levers

The simplest machine is a *lever*, which consists of two pieces or components: a *bar* (or beam) and a *fulcrum* (the pivot-point around which motion takes place). As shown below, the *effort* acts at a distance (L_1) from the fulcrum and the *load* acts at a distance (L_2) from the fulcrum.

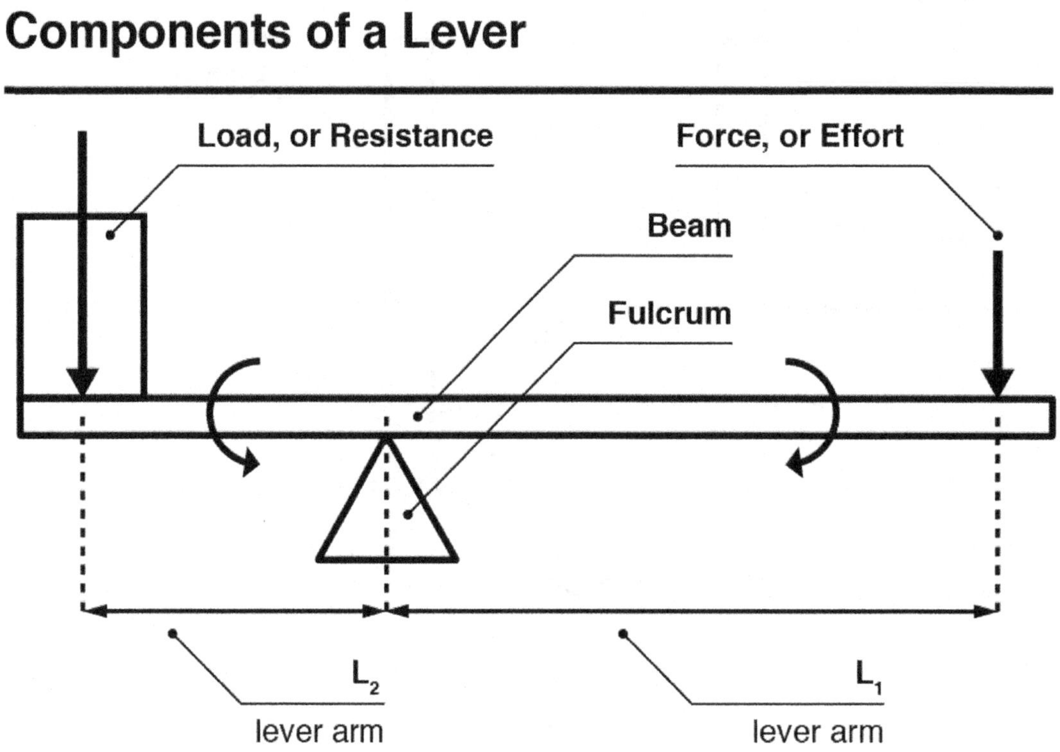

These lengths L_1 and L_2 are called *lever arms*. When the lever is balanced, the load (R) times its lever arm (L_2) equals the effort (F) times its lever arm (L_1). The force needed to lift the load is:

$$F = R \times \frac{L_2}{L_1}$$

This equation shows that as the lever arm L_1 is increased, the force required to lift the resisting load (R) is reduced. This is why Archimedes, one of the leading ancient Greek scientists, said, "Give me a lever long enough, and a place to stand, and I can move the Earth."

The ratio of the moment arms is the so-called "mechanical advantage" of the simple lever; the effort is multiplied by the mechanical advantage. For example, a 100-kilogram mass (a weight of approximately 1000 N) is lifted with a lever like the one in the figure below, with a total length of 3 meters, and the fulcrum situated 50 centimeters from the left end. What's the force needed to balance the load?

$$F = 1000\ N \times \frac{0.5\ meters}{2.5\ meters} = 200\ N$$

Depending on the location of the load and effort with respect to the fulcrum, three "classes" of lever are recognized. In each case, the forces can be analyzed as described above.

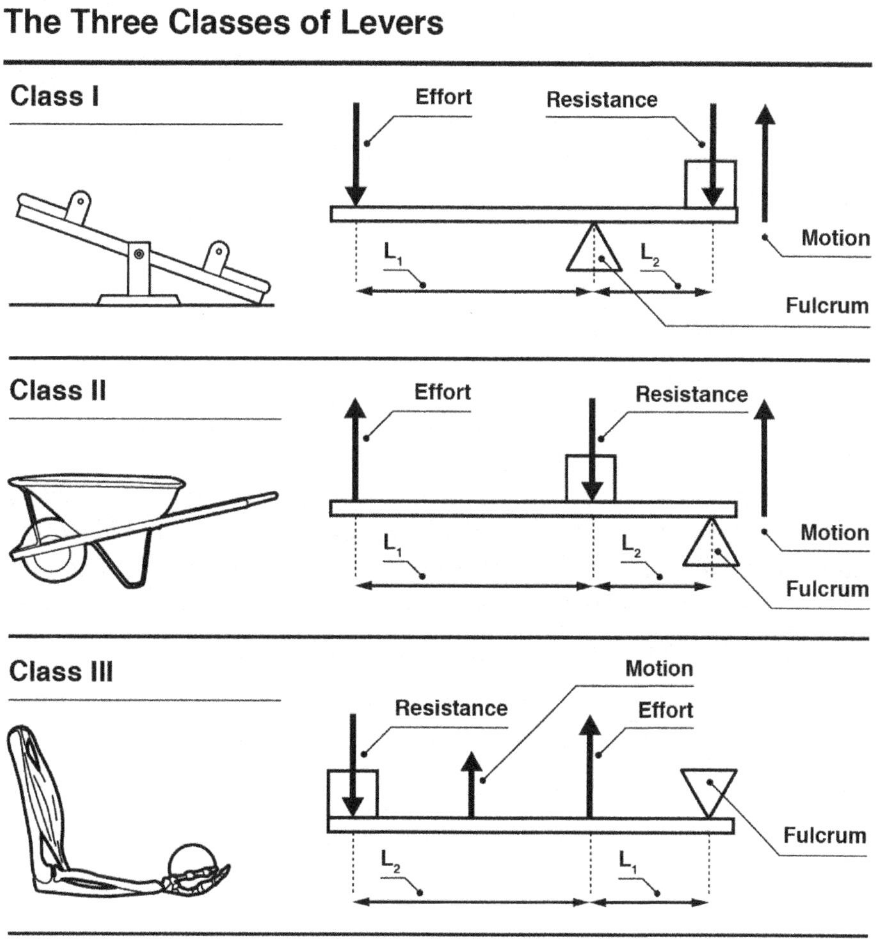

As seen in the figure, a *Class I* lever has the fulcrum positioned between the effort and the load. Examples of Class I levers include see-saws, balance scales, crow bars, and scissors. As explained above, the force needed to balance the load is $F = R \times (L_2/L_1)$, which means that the mechanical advantage is L_2/L_1. The crane boom shown back in the first figure in this section was a Class I lever, where the tower acted as the fulcrum and the counterweight on the left end of the boom provided the effort.

For a *Class II* lever, the load is placed between the fulcrum and the effort. A wheel barrow is a good example of a Class II lever. The mechanical advantage of a Class II lever is $(L_1 + L_2)/L_2$.

For a *Class III* lever, the effort is applied at a point between the fulcrum and the load, which increases the speed at which the load is moved. A human arm is a Class III lever, with the elbow acting as the fulcrum. The mechanical advantage of a Class III lever is $(L_1 + L_2)/L_1$.

Wheels and Axles

The wheel and axle is a special kind of lever. The *axle*, to which the load is applied, is set perpendicular to the *wheel* through its center. Effort is then applied along the rim of the wheel, either with a cable running around the perimeter or with a *crank* set parallel to the axle.

The mechanical advantage of the wheel and axle is provided by the moment arm of the perimeter cable or crank. Using the center of the axle (with a radius of *r*) as the fulcrum, the resistance of the load (*L*) is just balanced by the effort (*F*) times the wheel radius:

$$F \times R = L \times r \quad \text{or} \quad F = L \times \frac{r}{R}$$

This equation shows that increasing the wheel's radius for a given shaft reduces the required effort to carry the load. Of course, the axle must be made of a strong material or it'll be twisted apart by the applied torque. This is why steel axles are used.

Gears, Belts, and Cams

The functioning of a wheel and axle can be modified with the use of gears and belts. *Gears* are used to change the direction or speed of a wheel's motion.

The direction of a wheel's motion can be changed by using *beveled gears*, with the shafts set at right angles to each other, as shown in part *A* in the figure below.

The speed of a wheel can be changed by meshing together *spur gears* with different diameters. A small gear (A) is shown driving a larger gear (B) in the middle section *(B)* in the figure below. The gears rotate in opposite directions; if the driver, Gear A, moves clockwise, then Gear B is driven counter-clockwise. Gear B rotates at half the speed of the driver, Gear A. In general, the change in speed is given by the ratio of the number of teeth in each gear:

$$\frac{Rev_{Gear\ B}}{Rev_{Gear\ A}} = \frac{Number\ of\ Teeth\ in\ A}{Number\ of\ Teeth\ in\ B}$$

Rather than meshing the gears, *belts* are used to connect them as shown in part *(C)*.

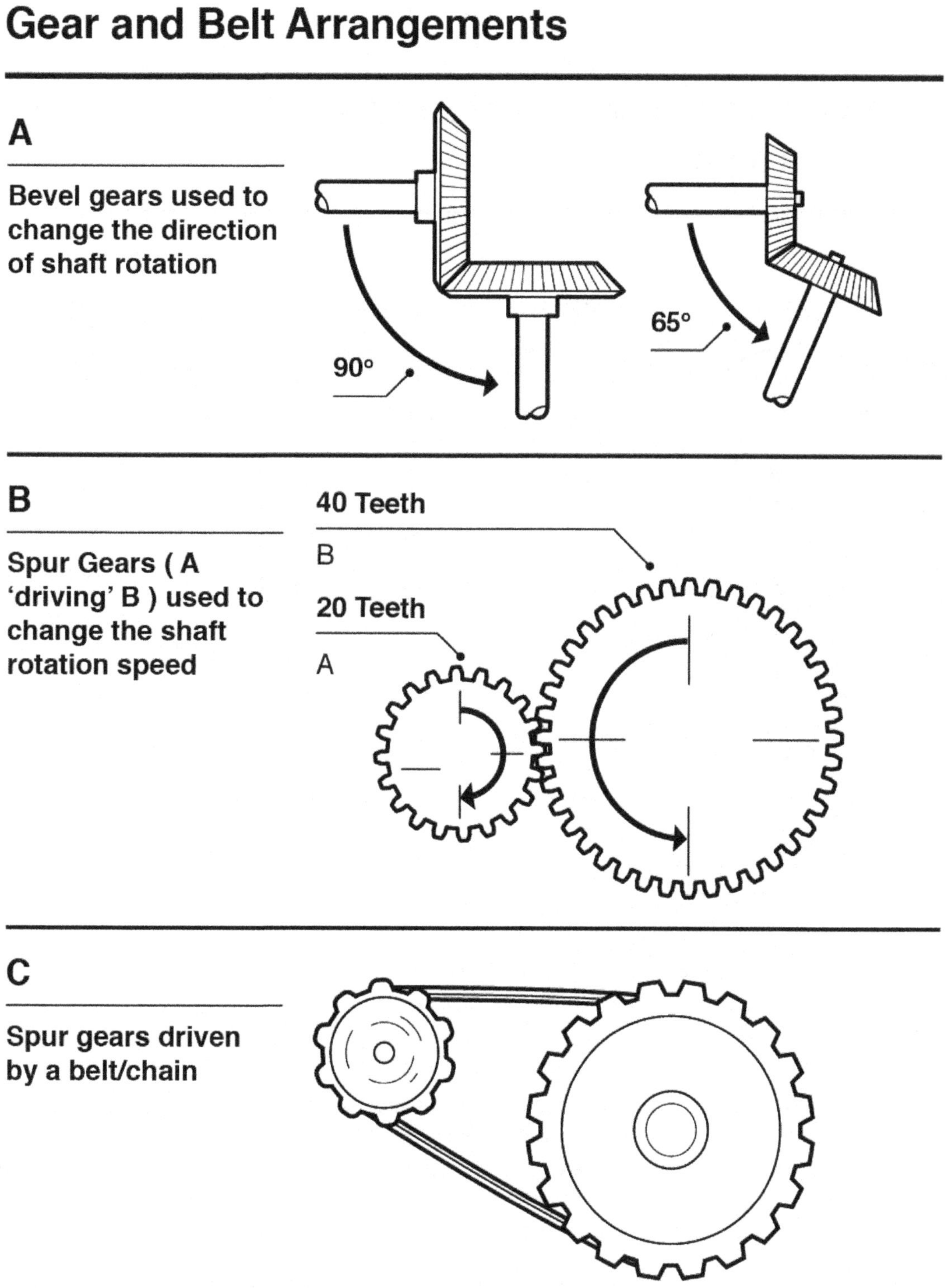

Gears can change the speed and direction of the axle rotation, but the rotary motion is maintained. To convert the rotary motion of a gear train into linear motion, it's necessary to use a *cam* (a type of off-centered wheel shown in the figure below, where rotary shaft motion lifts the valve in a vertical direction.

Conversion of Rotary to Vertical Linear Motion with a Cam

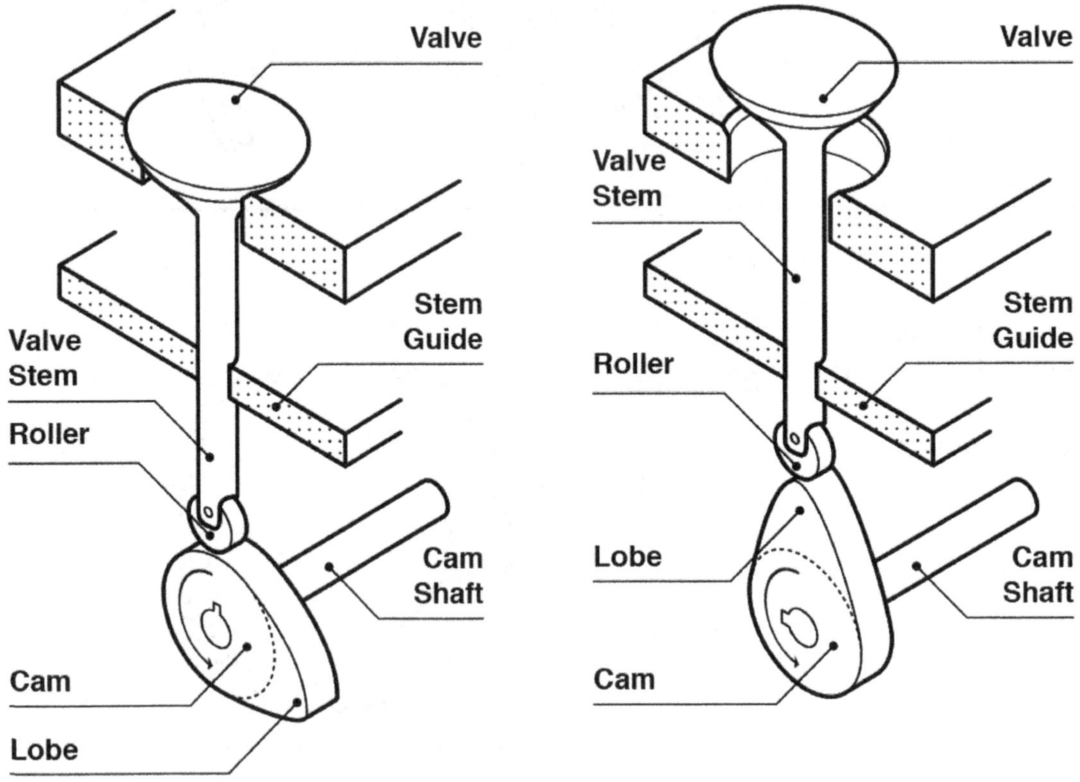

Pulleys

A *pulley* looks like a wheel and axle, but provides a mechanical advantage in a different way. A *fixed pulley* was shown previously as a way to capture the potential energy of a falling weight to do useful work by lifting another weight. As shown in part *A* in the figure below, the fixed pulley is used to change the direction of the downward force exerted by the falling weight, but it doesn't provide any mechanical advantage.

The lever arm of the falling weight (A) is the distance between the rim of the fixed pulley and the center of the axle. This is also the length of the lever arm acting on the rising weight (B), so the ratio of the two arms is 1:0, meaning there's no mechanical advantage. In the case of a wheel and axle, the mechanical advantage is the ratio of the wheel radius to the axle radius.

A *moving pulley*, which is really a Class II lever, provides a mechanical advantage of 2:1 as shown below on the right side of the figure *(B)*.

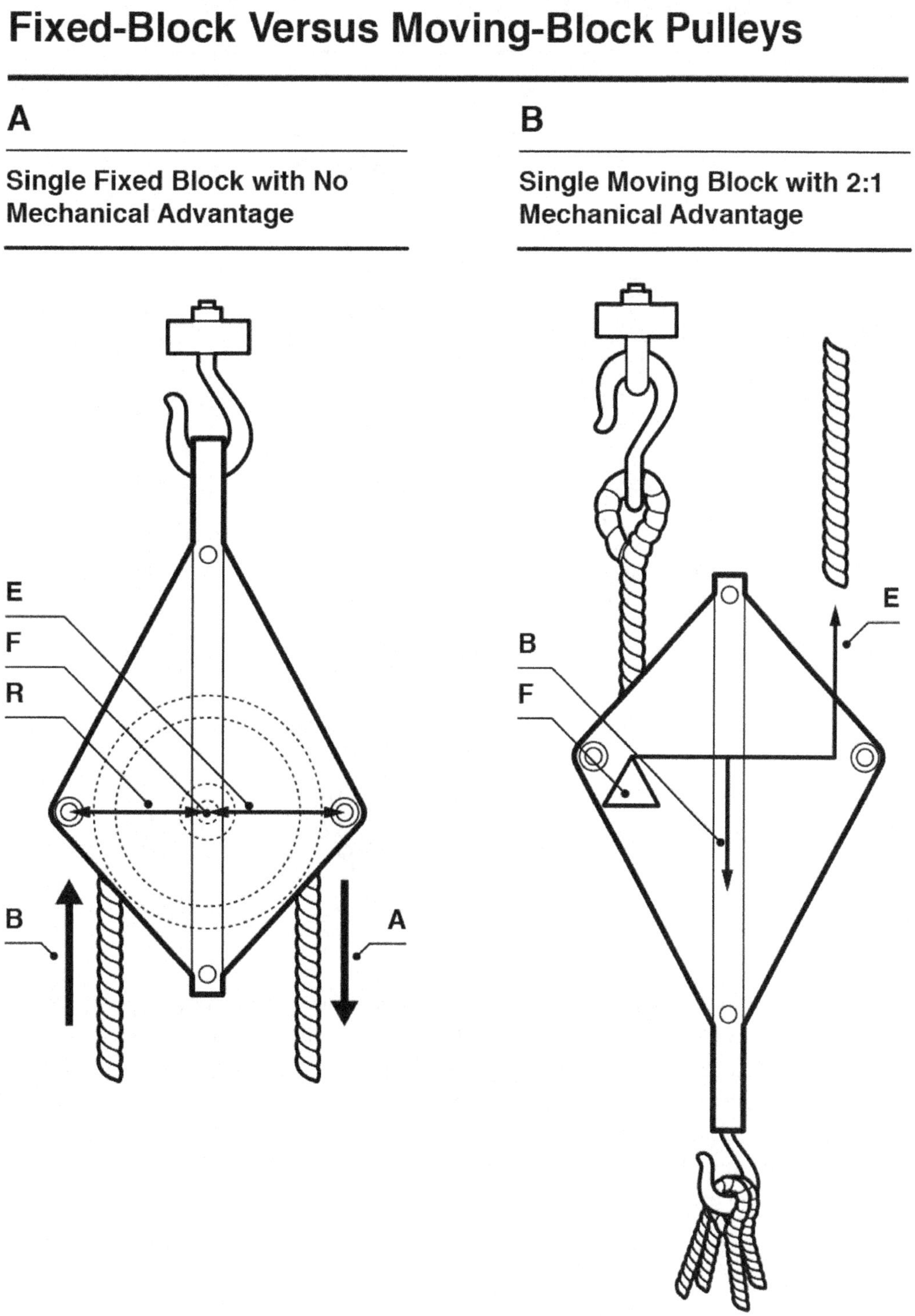

As demonstrated by the rigs in the figure below, using a wider moving block with multiple sheaves can achieve a greater mechanical advantage.

Single-Acting and Double-Acting Block and Tackles

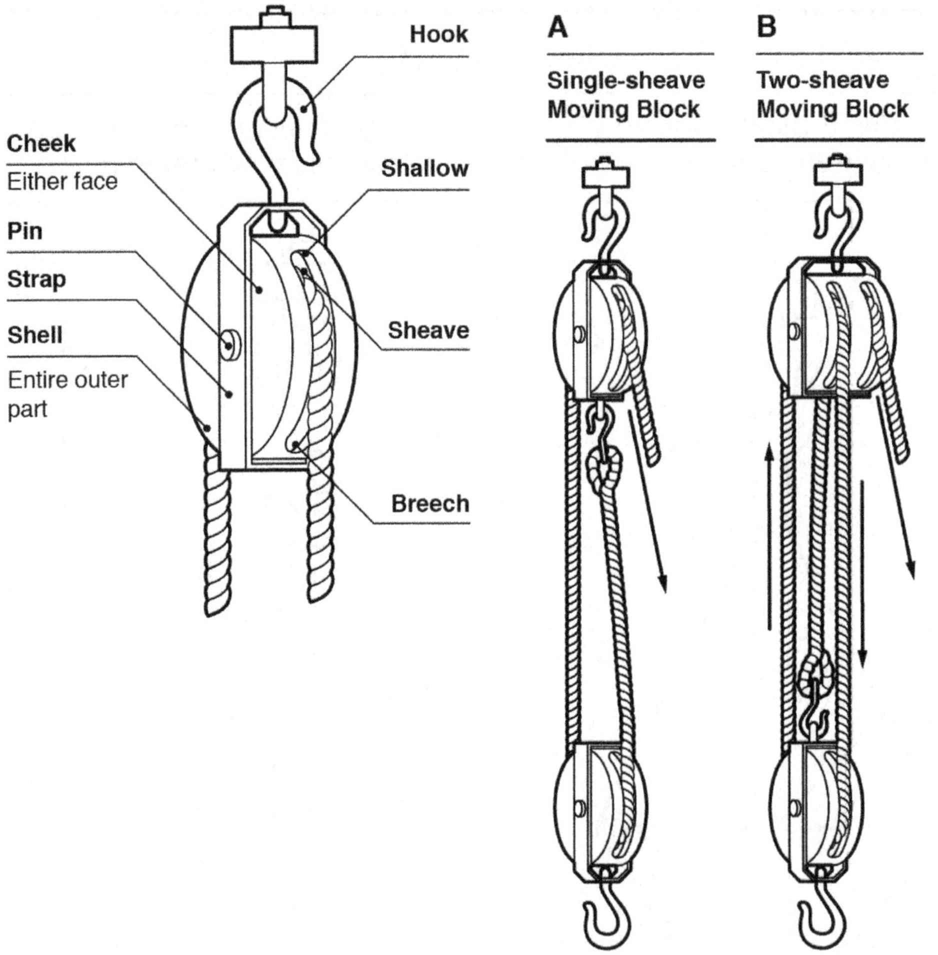

The mechanical advantage of the multiple-sheave block and tackle is approximated by counting the number of ropes going to and from the moving block. For example, there are two ropes connecting the moving block to the fixed block in part *A* of the figure above, so the mechanical advantage is 2:1. There are three ropes connecting the moving and fixed blocks in part *B*, so the mechanical advantage is 3:1. The advantage of using a multiple-sheave block is the increased hauling power obtained, but there's a cost; the weight of the moving block must be overcome, and a multiple-sheave block is significantly heavier than a single-sheave block.

Ramps

The *ramp* (or inclined plane) has been used since ancient times to move massive, extremely heavy objects up to higher positions, such as in the pyramids of the Middle East and Central America.

For example, to lift a barrel straight up to a height (*H*) requires a force equal to its weight (*W*). However, the force needed to lift the barrel is reduced by rolling it up a ramp, as shown below. So, if the ramp is *D* meters long and *H* meters high, the force (*F*) required to roll the weight (*W*) up the ramp is:

$$F = \frac{H}{D} \times W$$

Definition Sketch for a Ramp or Inclined Plane

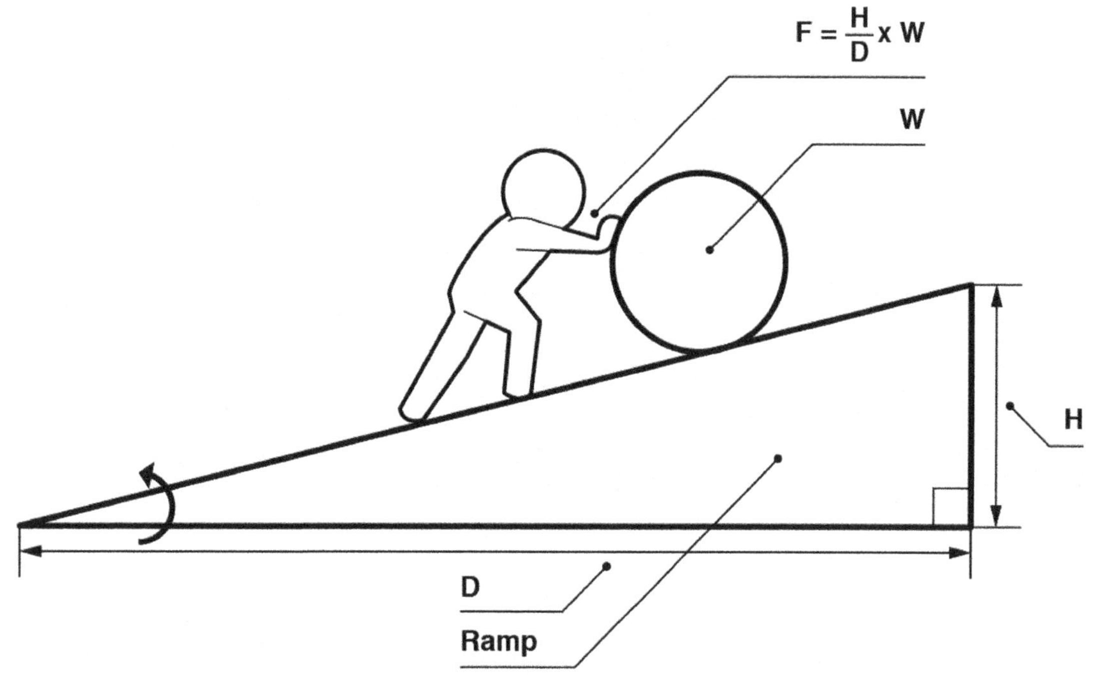

For a fixed height and weight, the longer the ramp, the less force must be applied. Remember, though, that the useful work done (in *N-m*) is the same in either case and is equal to *W* × *H*.

Wedges

If an incline or ramp is imagined as a right triangle like in the figure above, then a *wedge* would be formed by placing two inclines (ramps) back to back (or an isosceles triangle). A wedge is one of the six simple machines and is used to cut or split material. It does this by being driven for its full length into the material being cut. This material is then forced apart by a distance equal to the base of the wedge. Axes, chisels, and knives work on the same principle.

Screws

Screws are used in many applications, including vises and jacks. They are also used to fasten wood and other materials together. A screw is thought of as an inclined plane wrapped around a central cylinder. To visualize this, one can think of a barbershop pole, or cutting the shape of an incline (right triangle) out of a sheet of paper and wrapping it around a pencil (as in part *A* in the figure below). Threads are made from steel by turning round stock on a lathe and slowly advancing a cutting tool (a wedge) along it, as shown in part *B*.

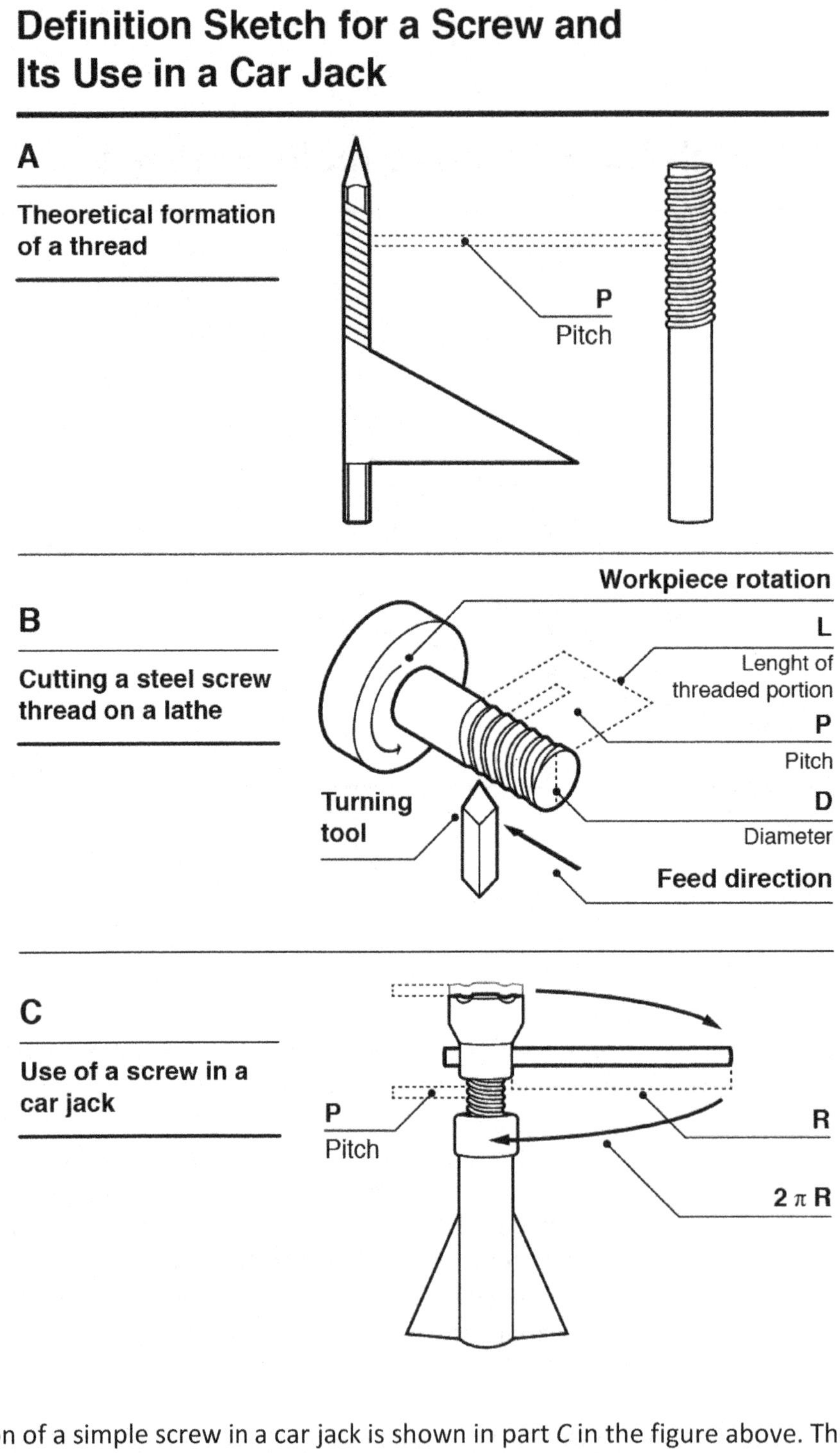

The application of a simple screw in a car jack is shown in part *C* in the figure above. The mechanical advantage of the jack is derived from the pitch of the screw winding. Turning the handle of the jack one revolution raises the screw by a height equal to the *screw pitch (p)*. If the handle has a length *R*, the distance the handle travels is equal to the circumference of the circle it traces out. The theoretical mechanical advantage of the jack's screw is:

$$MA = \frac{F}{L} = \frac{p}{2\pi R} \text{ so } \quad F = L \times \frac{p}{2\pi R}$$

For example, the theoretical force (*F*) required to lift a car with a mass (*L*) of 5000 kilograms, using a jack with a handle 30 centimeters long and a screw pitch of 0.5 cm, is given as:

$$F \cong 50{,}000\ N \times \frac{0.5\ cm}{6.284 * 30\ cm} \cong 130\ N$$

The theoretical value of mechanical advantage doesn't account for friction, so the actual force needed to turn the handle is higher than calculated.

Practice Questions

The following Practice Test contains sample problems that reinforce the principles presented in the *Mechanical Comprehension (MC)* study guide. The answers to these problems, along with a brief explanation, follows.

1. A car is traveling at a constant velocity of 25 m/s. How long does it take the car to travel 45 kilometers in a straight line?
   a. 1 hour
   b. 3600 seconds
   c. 1800 seconds
   d. 900 seconds
   e. 4500 seconds

2. A ship is traveling due east at a speed of 1 m/s against a current flowing due west at a speed of 0.5 m/s. How far has the ship travelled from its point of departure after two hours?
   a. 1.8 kilometers west of its point of departure
   b. 3.6 kilometers west of its point of departure
   c. 1.8 kilometers east of its point of departure
   d. 3.6 kilometers east of its point of departure
   e. 1.0 kilometers west of its point of departure

3. A car is driving along a straight stretch of highway at a constant speed of 60 km/hour when the driver slams the gas pedal to the floor, reaching a speed of 132 km/hour in 10 seconds. What's the average acceleration of the car after the engine is floored?
   a. 1 m/s^2
   b. 2 m/s^2
   c. 3 m/s^2
   d. 4 m/s^2
   e. 5 m/s^2

4. A spaceship with a mass of 100,000 kilograms is far away from any planet. To accelerate the craft at the rate of 0.5 m/sec^2, what is the rocket thrust?
   a. 98.1 N
   b. 25,000 N
   c. 50,000 N
   d. 75,000 N
   e. 100,000 N

5. The gravitational acceleration on Earth averages 9.81 m/s^2. An astronaut weighs 1962 N on Earth. The diameter of Earth is six times the diameter of its moon. What's the mass of the astronaut on Earth's moon?
   a. 50 kilograms
   b. 100 kilograms
   c. 200 kilograms
   d. 300 kilograms
   e. 400 kilograms

6. A football is kicked so that it leaves the punter's toe at a horizontal angle of 45 degrees. Ignoring any spin or tumbling, at what point is the upward vertical velocity of the football at a maximum?

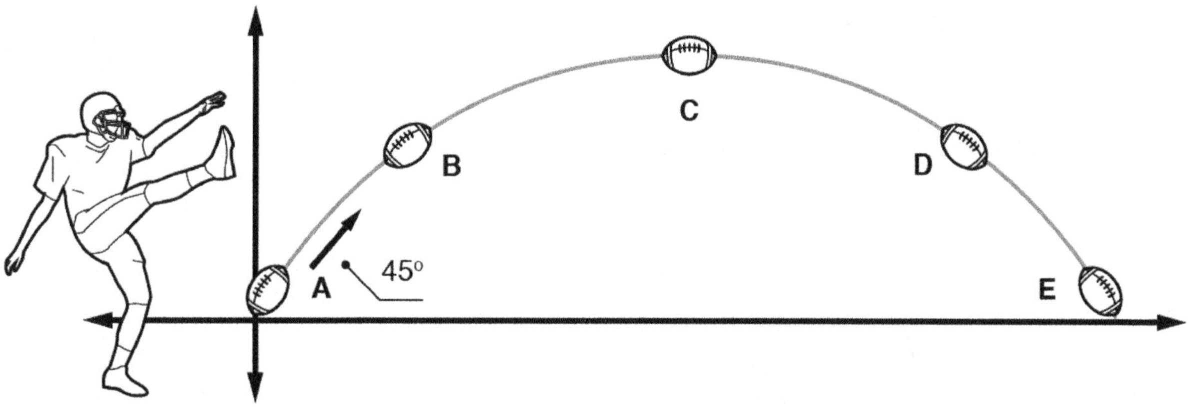

a. At Point A
b. At Point C
c. At Points B and D
d. At Points A and E
e. At Points A, C, and E

7. The skater is shown spinning in Figure (a), then bringing in her arms in Figure (b). Which sequence accurately describes what happens to her angular velocity?

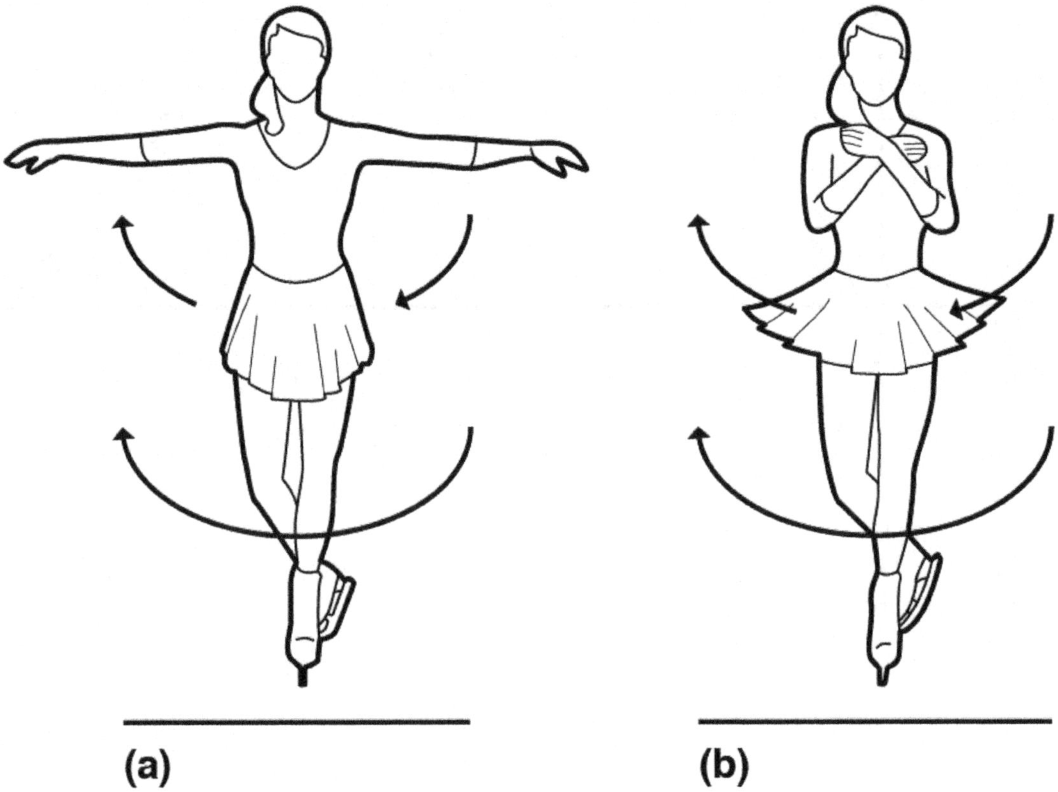

a. Her angular velocity decreases from (a) to (b)
b. Her angular velocity doesn't change from (a) to (b)
c. Her angular velocity increases from (a) to (b)
d. Her change in angular velocity in both (a) and (b) depends on where she is in the turn.
e. It's not possible to determine what happens to her angular velocity if her weight is unknown.

8. A cannonball is dropped from a height of 10 meters off of the ground. What is its approximate velocity just before it hits the ground?
a. 9.81 m/s
b. 14 m/s
c. 32 m/s
d. 1 m/s
e. It can't be determined without knowing the cannonball's mass

9. The pendulum is held at point A, and then released to swing to the right. At what point does the pendulum have the greatest kinetic energy?

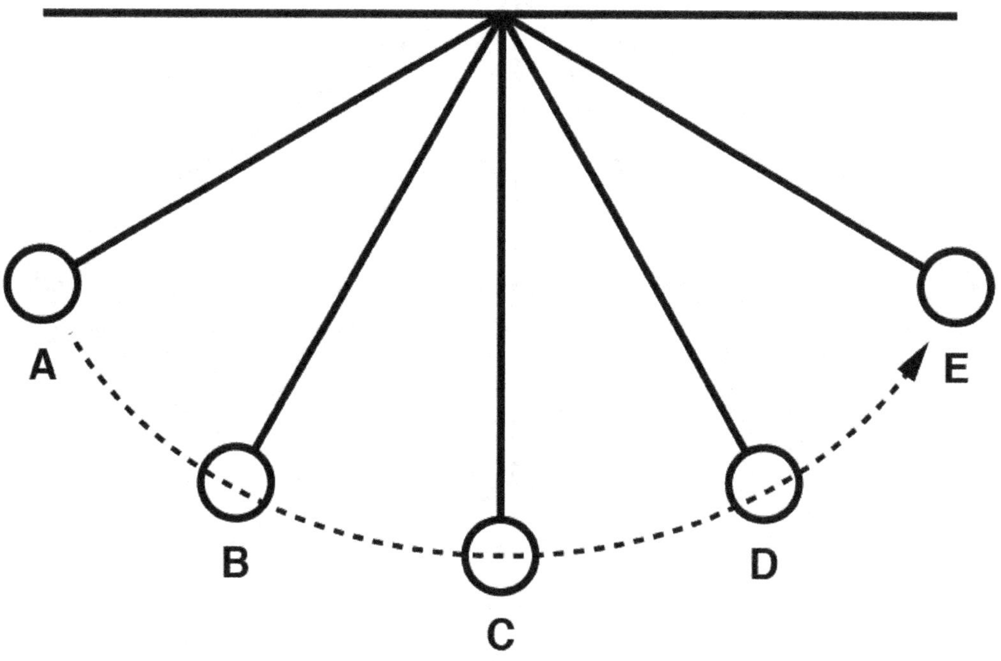

a. At Point A
b. At Point B
c. At Point C
d. At Point D
e. At Point E

10. Which statement is true of the total energy of the pendulum?

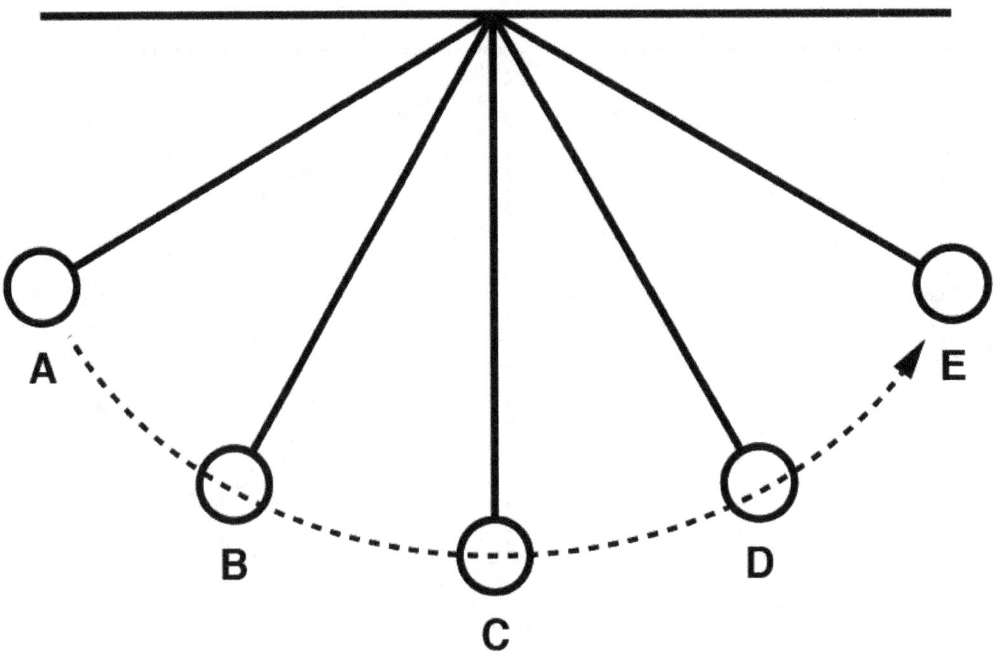

a. The total energy can't be determined without knowing the pendulum's mass.
b. Its total energy is at a maximum and equal at Points A and E.
c. Its total energy is at a maximum at Point C.
d. Its total energy is at a minimum at Point C.
e. Its total energy is the same at Points A, B, C, D, and E.

11. How do you calculate the useful work performed in lifting a 10-kilogram weight from the ground to the top of a 2-meter ladder?

a. 10kg x 2m x 32 m/s^2
b. 10kg x $2m^2$ x 9.81 m/s
c. 10kg x 2m x $9.81m/s^2$
c. 10kg x $2m^2$ x $9.81m/s^2$
e. It can't be determined without knowing the ground elevation

12. A steel spring is loaded with a 10-newton weight and is stretched by 0.5 centimeters. What is the deflection if it's loaded with two 10-newton weights?

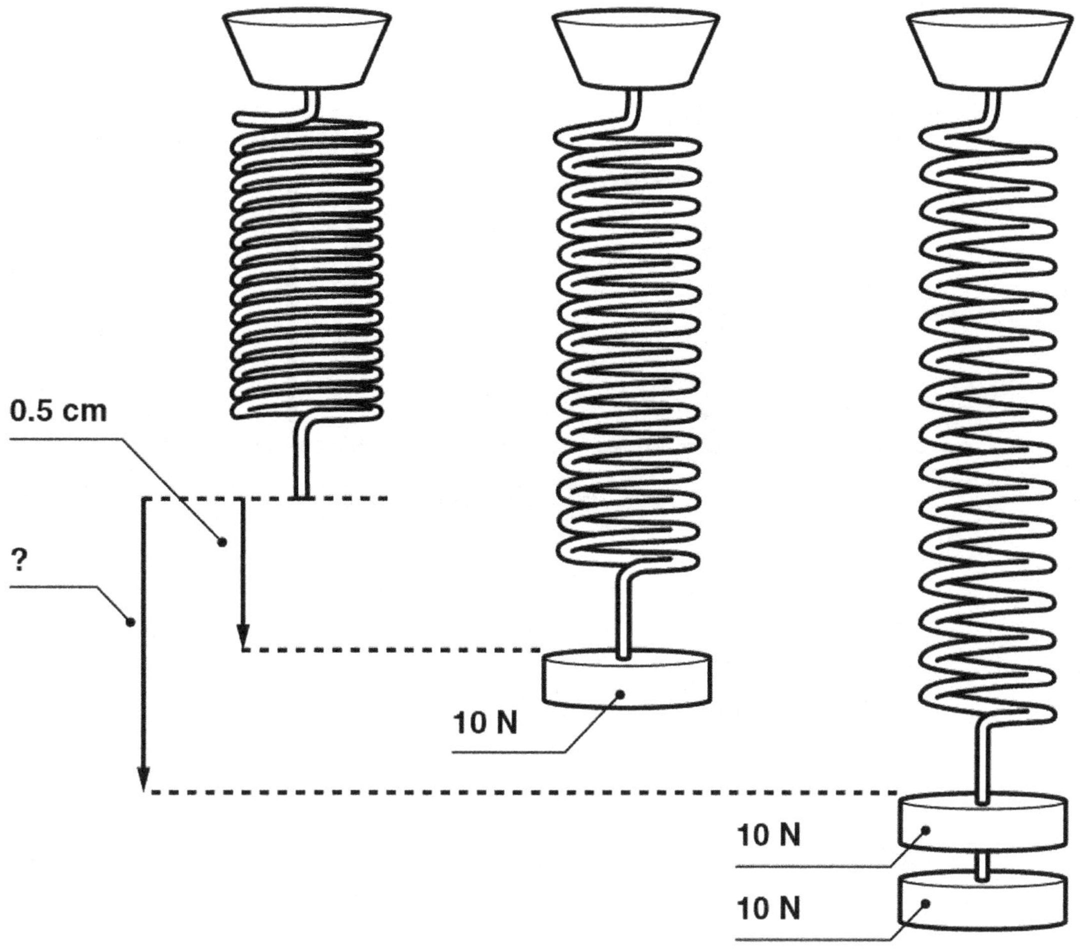

a. 0.5 centimeter
b. 1 centimeter
c. 2 centimeters
d. 10 centimeters
e. It can't be determined without knowing the Modulus of Elasticity of the steel.

13. A 1000-kilogram concrete block is resting on a wooden surface. Between these two materials, the coefficient of sliding friction is 0.4 and the coefficient of static friction is 0.5. How much more force is needed to get the block moving than to keep it moving?

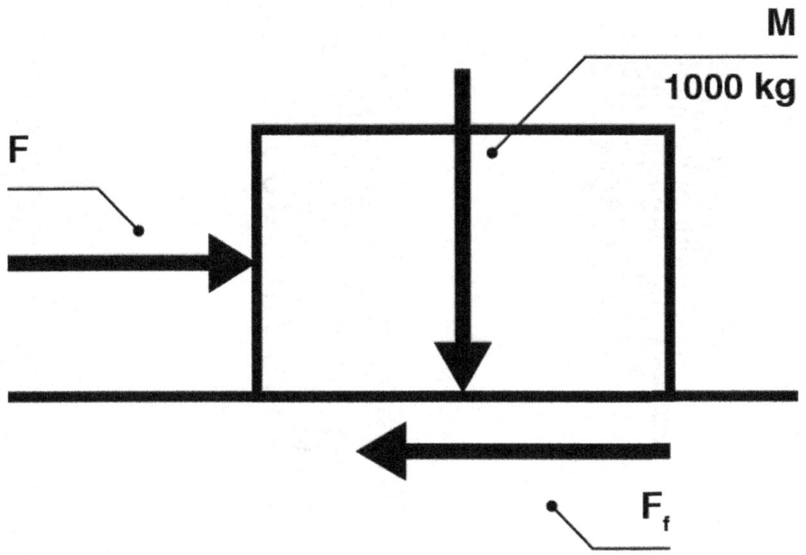

a. 981 N
b. 1962 N
c. 3924 N
d. 9810 N
e. 10000 N

14. The master cylinder (F_1) of a hydraulic jack has a cross-sectional area of 0.1 m^2, and a force of 50 N is applied. What must the area of the drive cylinder (F_2) be to support a weight of 800 N?

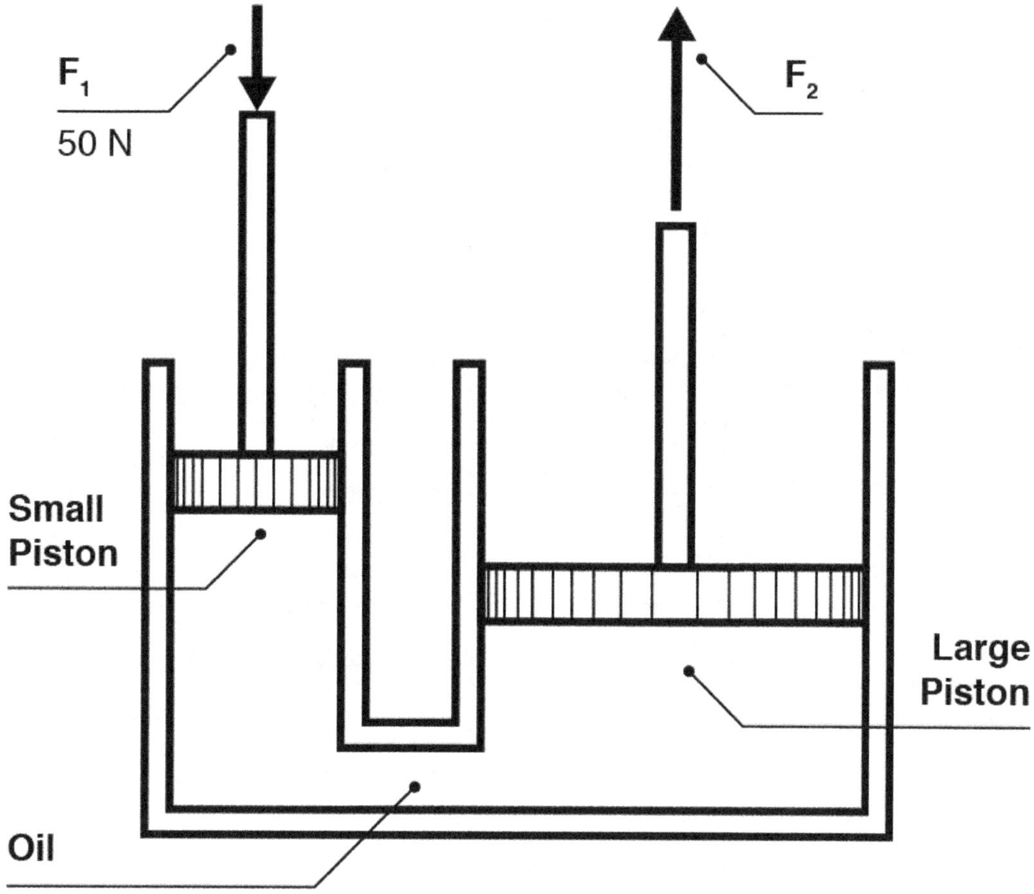

a. 0.1 m^2
b. 0.4 m^2
c. 0.8 m^2
d. 1.6 m^2
e. 3.2 m^2

15. A gas with a volume V_1 is held down by a piston with a force of *F* newtons. The piston has an area of *A*. After heating the gas, it expands against the weight to a volume V_2. What was the work done?

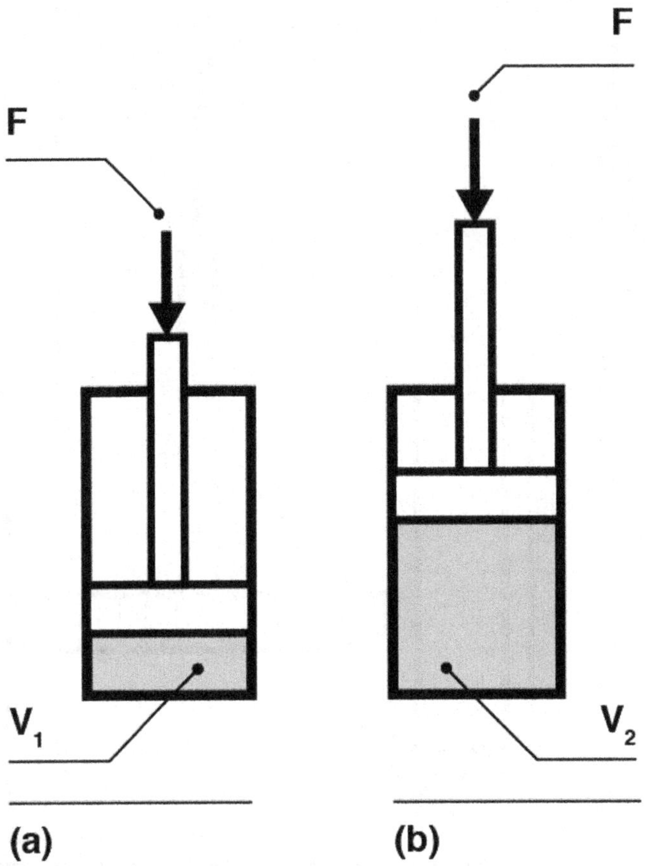

a. *F/A*
b. $(F/A) \times V_1$
c. $(F/A) \times V_2$
d. $(F/A) \times (V_1 \times V_2)$
e. $(F/A) \times (V_2 - V_1)$

16. A 1000-kilogram weight is raised 30 meters in 10 minutes. What is the approximate power expended in the period?
a. 1000 Kg × m/s^2
b. 500 N·m
c. 500 Kg x m
d. 100 watts
e. 500 J/s

17. A 2-meter high, concrete block is submerged in a body of water 12 meters deep (as shown). Assuming that the water has a unit weight of 1000 N/m^3, what is the pressure acting on the upper surface of the block?

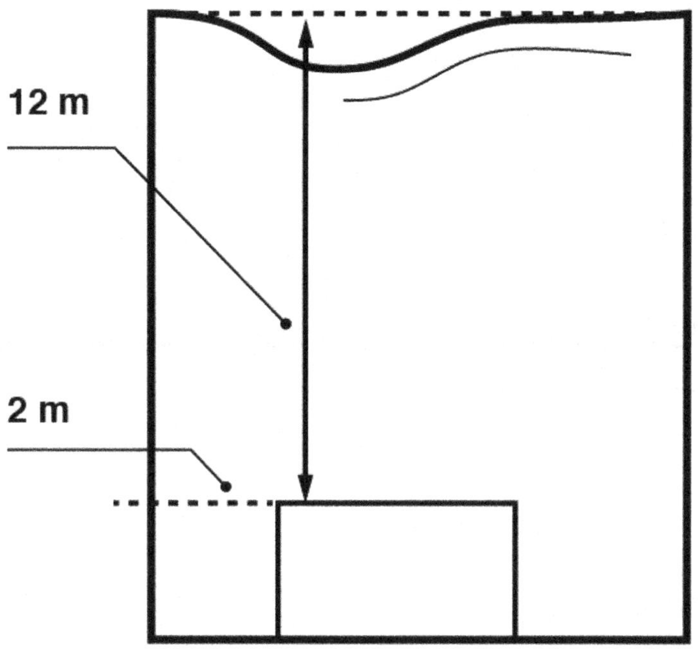

a. 2,000 Pa
b. 10,000 Pa
c. 12,000 Pa
d. 14,000 Pa
e. It can't be calculated without knowing the top area of the block.

18. Closed Basins A and B each contain a 10,000-ton block of ice. The ice block in Basin A is floating in sea water. The ice block in Basin B is aground on a rock ledge (as shown). When all the ice melts, what happens to the water level in Basin A and Basin B?

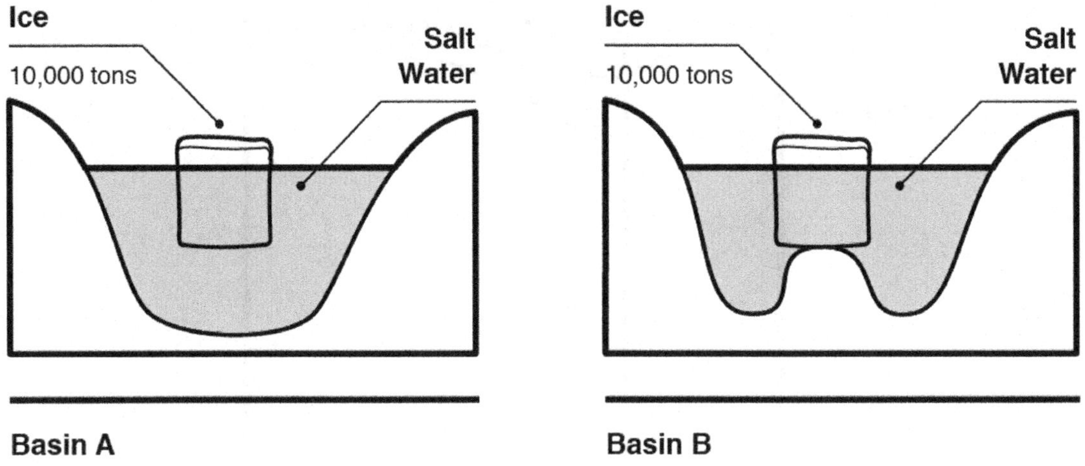

a. Water level rises in A but not in B
b. Water level rises in B but not in A
c. Water level rises in neither A nor B
d. Water level rises in both A and B
e. It cannot be determined without knowing the temperature

19. An official 10-lane Olympic pool is 50 meters wide by 25 meters long. How long does it take to fill the pool to the recommended depth of 3 meters using a pump with a 750 liter per second capacity?
a. 750 seconds
b. 2500 seconds
c. 5000 seconds
d. 10,000 seconds
e. 100,000 seconds

20. Water is flowing in a rectangular canal 10 meters wide by 2 meters deep at a velocity of 3 m/s. The canal is half full. What is the flow rate?
a. 20 m^3/s
b. 30 m^3/s
c. 60 m^3/s
d. 90 m^3/s
e. 120 m^3/s

21. A 150-kilogram mass is placed on the left side of the lever as shown. What force must be exerted on the right side (in the location shown by the arrow) to balance the weight of this mass?

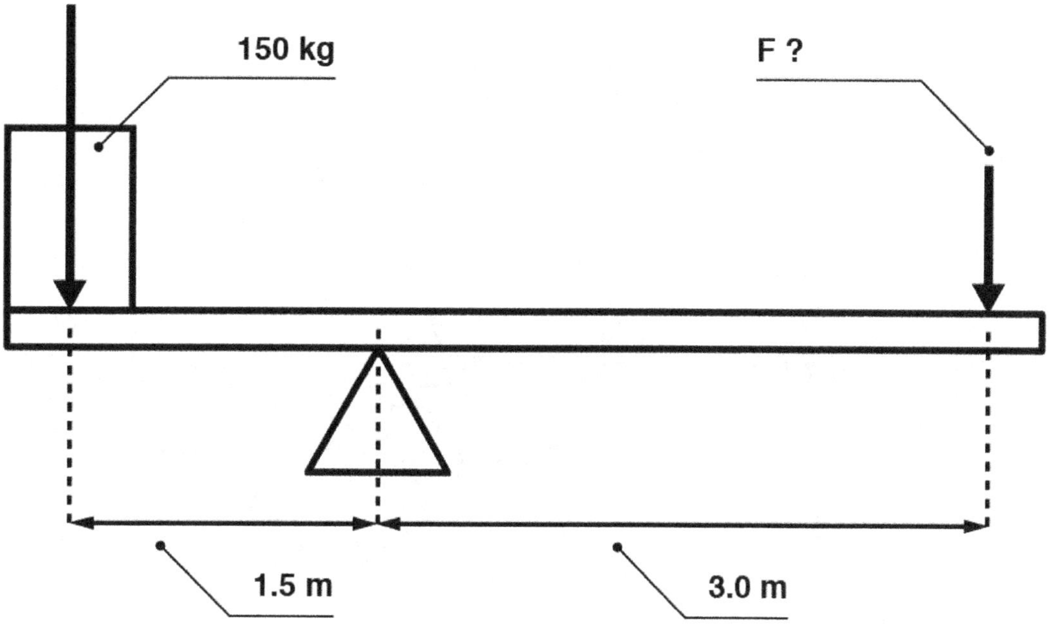

a. 350 kg·m
b. 675 kg·m
c. 737.75 N
d. 1471.5 N
e. 2207.25 N·m

22. For the wheel and axle assembly shown, the shaft radius is 20 millimeters and the wheel radius is 300 millimeters. What's the required effort to lift a 600 N load?

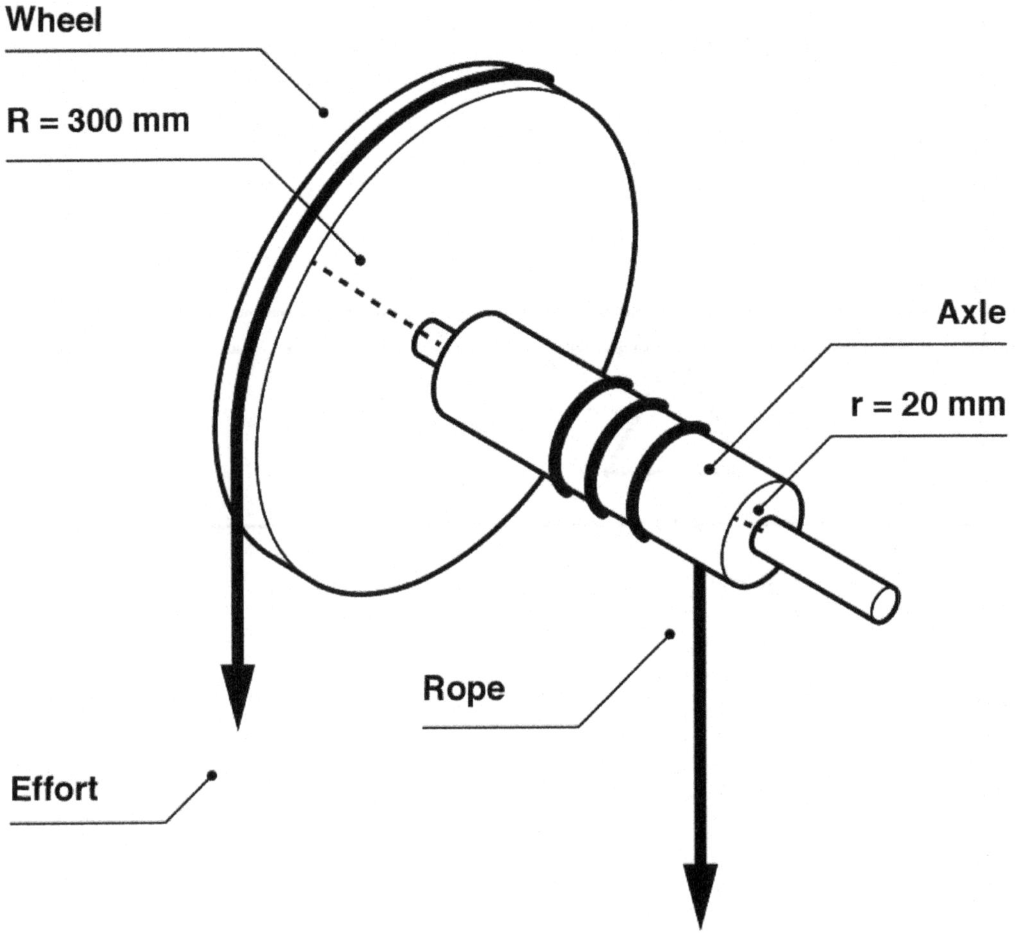

a. 10 N
b. 20 N
c. 30 N
d. 40 N
e. 50 N

23. The driver gear (Gear A) turns clockwise at a rate of 60 RPM. In what direction does Gear B turn and at what rotational speed?

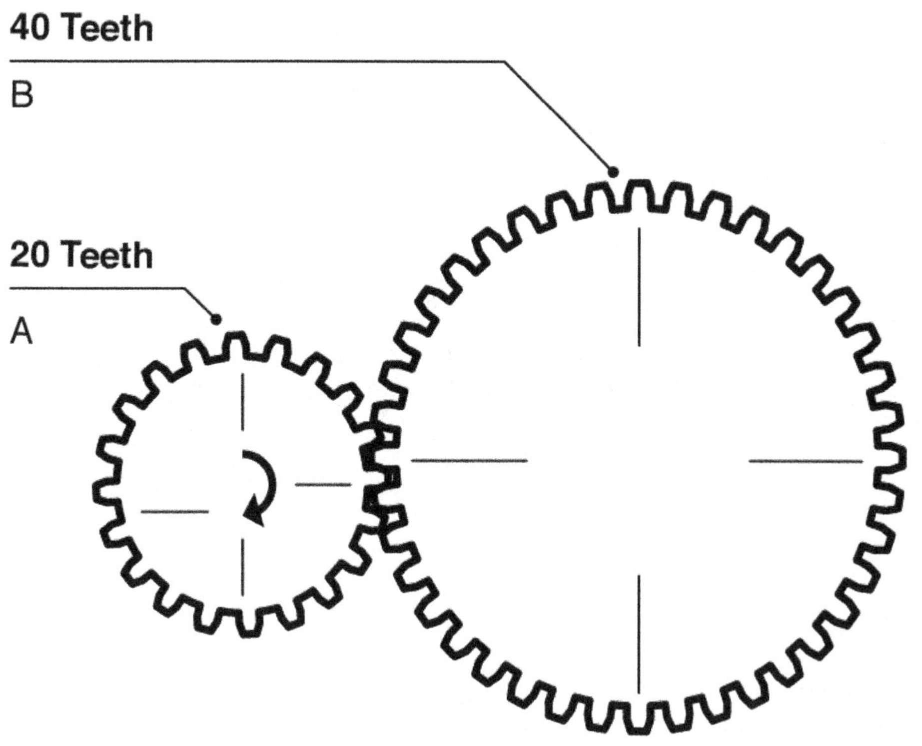

a. Clockwise at 120 RPM
b. Counterclockwise at 120 RPM
c. Clockwise at 30 RPM
d. Clockwise at 60 RPM
e. Counterclockwise at 30 RPM

24. The three steel wheels shown are connected by rubber belts. The two wheels at the top have the same diameter, while the wheel below is twice their diameter. If the driver wheel at the upper left is turning clockwise at 60 RPM, at what speed and in which direction is the large bottom wheel turning?

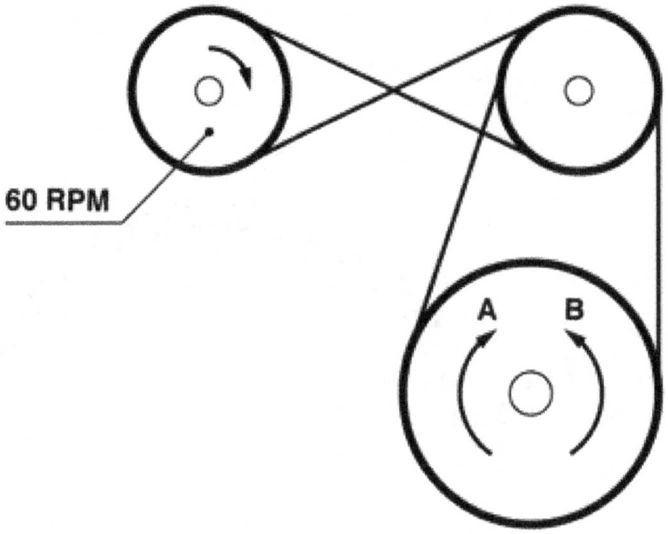

a. 30 RPM, clockwise (A)
b. 30 RPM, counterclockwise (B)
c. 120 RPM, clockwise (A)
d. 120 RPM, counterclockwise (B)
e. It cannot be determined without knowing the length of the rope.

25. In case (a), both blocks are fixed. In case (b), the load is hung from a moveable block. Ignoring friction, what is the required force to move the blocks in both cases?

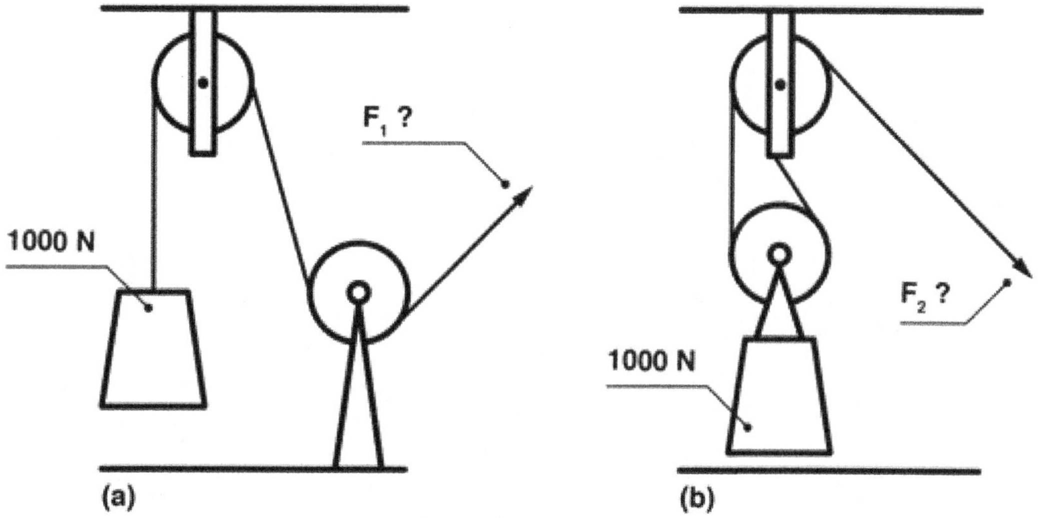

a. F_1 = 500 N; F_2 = 500 N
b. F_1 = 500 N; F_2 = 1000 N
c. F_1 = 1000 N; F_2 = 500 N
d. F_1 = 1000 N; F_2 = 1000 N
e. F_1 = 1000 N; F_2 = 2000 N

Answer Explanations

1. C: The answer is 1800 seconds:

$$\left(Desired\ Distance\ in\ km\ \times\ conversion\ factor\ (m\ to\ km)\right)/current\ velocity\ in \frac{m}{s}$$

$$\left(45\ km\ \times \frac{1000\ m}{km}\right)/25\frac{m}{s} = 1800\ seconds$$

2. D: The answer is 3.6 kilometers east of its point of departure. The ship is traveling faster than the current, so it will be east of the starting location. Its net forward velocity is 0.5 m/s, which is 1.8 kilometers/hour, or 3.6 kilometers in two hours.

3. B: The answer is 2 m/s^2:

$$a = \frac{\Delta v}{\Delta t} = \frac{132\frac{km}{hr} - 60\frac{km}{hr}}{10\ seconds}$$

$$\frac{70\frac{km}{hr} \times 1000\frac{m}{km} \times \frac{hour}{3600\ sec}}{10\ seconds} = 2\ m/s^2$$

4. C: The answer is 50,000 N. The equation $F = ma$ should be memorized. All of the values are given in the correct units (kilogram-meter-second) so just plug them in.

5. C: The answer is 200 kilograms. This is actually a trick question. The mass of the astronaut is the same everywhere (it is the weight that varies from planet to planet). The astronaut's mass in kilograms is calculated by dividing his weight on Earth by the acceleration of gravity on Earth: 1962/9.81 = 200.

6. A: The answer is that the upward velocity is at a maximum when it leaves the punter's toe. The acceleration due to gravity reduces the upward velocity every moment thereafter. The speed is the same at points A and E, but the velocity is different. At point E, the velocity has a maximum *negative* value.

7. C: The answer is her angular velocity increases from (a) to (b) as she pulls her arms in close to her body and reduces her moment of inertia.

8. B: The answer is 14 m/s. Remember that the cannonball at rest "y" meters off the ground has a potential energy of $PE = mgy$. As it falls, the potential energy is converted to kinetic energy until (at ground level) the kinetic energy is equal to the total original potential energy:

$$\frac{1}{2}mv^2 = mgy, \text{ or } v = \sqrt{2gy}$$

This makes sense because all objects fall at the same rate, so the velocity *must* be independent of the mass (which is why Choice *E* is incorrect). Plugging the values into the equation, the result is 14 m/s. Remember, the way to figure this quickly is to have $g = 10$ rather than 9.81.

9. C: The answer is at Point C, the bottom of the arc.

10. E: This question isn't difficult, but it must be read carefully:

The total energy is conserved, so it's the same at *all* points on the arc. Choice *A* is wrong. The motion of a pendulum is independent of the mass. Just like how all objects fall at the same rate, all pendulum bobs swing at the same rate, dependent on the length of the cord.

Choice *B* is wrong. Even though the total energy is at a maximum at Points A and E, it isn't equal at only those points. The total energy is the same at *all* points. Choices *C* and *D* are wrong. The kinetic energy is at a maximum at C, but not the *total* energy. The correct answer is *E*.

11. C: The answer is 10kg x 2m x 9.81m/s^2. This is easy, but it must also be read carefully. Choice *E* is incorrect because it isn't necessary to know the ground elevation. The potential energy is measured *with respect* to the ground and the ground (or datum elevation) can be set to any arbitrary value.

12. B: The answer is 1 centimeter. Remember that the force (*F*) required to stretch a spring a certain amount (*d*) is given by the equation *F* = *kd*. Therefore, *k* = *F/d* = 20N/0.5 cm = 20 N/cm. Doubling the weight to 20 N gives the deflection:

$$d = \frac{F}{k} = \frac{20N}{20N/cm} = 1\ centimeter$$

All of the calculations can be bypassed by remembering that the relation between force and deflection is linear. This means that doubling the force doubles the deflection, as long as the spring isn't loaded past its elastic limit.

13. A: The answer is 981 N. The start-up and sliding friction forces are calculated in the same way: normal force (or weight) multiplied by the friction coefficient. The difference between the two coefficients is 0.1, so the difference in forces is 0.1 x 1000 x 9.81 = 981 N.

14. D: The answer is 1.6 m^2. The pressure created by the load is 50N/0.1m^2 = 500 N/m^2. This pressure acts throughout the jack, including the large cylinder. Force is pressure times area, so the area equals pressure divided by force or 800N/500N/m^2 = 1.6m^2.

15. E: The answer is (*F/A*) x (V_2 -V_1). Remember that the work for a piston expanding is pressure multiplied by change in volume. Pressure = *F/A*. Change in volume is (V_2 - V_1).

16. E: The answer is 500 J/s. Choice *A* is incorrect because kg x m/s^2 is an expression of force, not power. Choice *B* is incorrect because N·m is an expression of work, not power. Choice *C* also has incorrect units for power. That leaves Choices *D* and *E*, both of which are expressed in units of power: watts or joules/second. Using an approximate calculation (as suggested):

$$1000\ kg \times 10\frac{m}{s^2} \times 30\ m = 300{,}000\ N \cdot m \quad \text{so} \quad \frac{300{,}000\ N \cdot m}{600\ seconds} = 500\ watts = 500\ J/s$$

17. C: The answer is 12,000 Pa. The top of the block is under 12 meters of water:

$$P = 1000\frac{N}{m^3} \times 12\ meters = 12{,}000\frac{N}{m^2} = 12{,}000\ Pa$$

There are two "red herrings" here: Choice *D* of 14,000 Pa is the pressure acting on the *bottom* of the block (perhaps through the sand on the bottom of the bay). Choice *E* (that it can't be calculated without knowing the top area of the block) is also incorrect. The top area is needed to calculate the total *force* acting on the top of the block, not the pressure.

18. B: The answer is that the water level rises in B but not in A. Why? Because ice is not as dense as water, so a given mass of water has more volume in a solid state than in a liquid state. Thus, it floats. As the mass of ice in Basin A melts, its volume (as a liquid) is reduced. In the end, the water level doesn't change. The ice in Basin B isn't floating. It's perched on high ground in the center of the basin. When it melts, water is added to the basin and the water level rises.

19. C: The answer is 5000 seconds. The volume is 3 x 25 x 50 = 3750 m^3. The volume divided by the flow rate gives the time. Since the pump capacity is given in liters per second, it's easier to convert the volume to liters. One thousand liters equals a cubic meter:

$$Time = \frac{3{,}750{,}000\ liters}{750\ liters/second} = 5000\ seconds = 1.39\ hours$$

20. B: The answer is 30 m^3/s. One of the few equations that must be memorized is $Q = vA$. The area of flow is 1m x 10m because only half the depth of the channel is full of water.

21. C: The answer is 737.75 N. This is a simple calculation:

$$\frac{9.81\ m}{s^2} \times 150\ kg \times 1.5\ m = 3\ m \times F \quad \text{so} \quad F = \frac{2207.25\ N \cdot m}{3\ meters}$$

22. D: The answer is 40 N. Use the equation $F = L \times r/R$. Note that for an axle with a given, set radius, the larger the radius of the wheel, the greater the mechanical advantage.

23. E: The answer is counterclockwise at 30 RPM. The driver gear is turning clockwise, and the gear meshed with it turns counter to it. Because of the 2:1 gear ratio, every revolution of the driver gear causes half a revolution of the follower.

24. B: The answer is 30 RPM, counterclockwise (B). While meshed gears rotate in different directions, wheels linked by a belt turn in the same direction. This is true unless the belt is twisted, in which case they rotate in opposite directions. So, the twisted link between the upper two wheels causes the right-hand wheel to turn counterclockwise, and the bigger wheel at the bottom also rotates counterclockwise. Since it's twice as large as the upper wheel, it rotates with half the RPMs.

25. C: The answer is F_1 = 1000 N; F_2 = 500 N. In case (a), the fixed wheels only serve to change direction. They offer no mechanical advantage because the lever arm on each side of the axle is the same. In case (b), the lower moveable block provides a 2:1 mechanical advantage. A quick method for calculating the mechanical advantage is to count the number of lines supporting the moving block (there are two in this question). Note that there are no moving blocks in case (a).

Dear SIFT Test Taker,

We would like to start by thanking you for purchasing this study guide for your SIFT exam. We hope that we exceeded your expectations.

Our goal in creating this study guide was to cover all of the topics that you will see on the test. We also strove to make our practice questions as similar as possible to what you will encounter on test day. With that being said, if you found something that you feel was not up to your standards, please send us an email and let us know.

We would also like to let you know about other books in our catalog that may interest you.

Test Name	**Amazon Link**
ASVAB	amazon.com/dp/1628454970
AFOQT	amazon.com/dp/1628454776
OAR	amazon.com/dp/1628456388
ASTB	amazon.com/dp/1628456809

We have study guides in a wide variety of fields. If the one you are looking for isn't listed above, then try searching for it on Amazon or send us an email.

Thanks Again and Happy Testing!
Product Development Team
info@studyguideteam.com

Interested in buying more than 10 copies of our product? Contact us about bulk discounts:
bulkorders@studyguideteam.com

FREE Test Taking Tips DVD Offer

To help us better serve you, we have developed a Test Taking Tips DVD that we would like to give you for FREE. **This DVD covers world-class test taking tips that you can use to be even more successful when you are taking your test.**

All that we ask is that you email us your feedback about your study guide. Please let us know what you thought about it – whether that is good, bad or indifferent.

To get your **FREE Test Taking Tips DVD**, email freedvd@studyguideteam.com with "FREE DVD" in the subject line and the following information in the body of the email:

a. The title of your study guide.

b. Your product rating on a scale of 1-5, with 5 being the highest rating.

c. Your feedback about the study guide. What did you think of it?

d. Your full name and shipping address to send your free DVD.

If you have any questions or concerns, please don't hesitate to contact us at freedvd@studyguideteam.com.

Thanks again!

Made in the USA
Monee, IL
10 February 2020

21592853R00131